Lineberger Memorial
Library
Lutheran Theological Southern Seminary Columbia, S. C.

BIO-BABEL

BIO—BABEL

Can We Survive the New Biology?

ALLEN R. UTKE

JOHN KNOX PRESS
ATLANTA

Library of Congress Cataloging in Publication Data
Utke, Allen R 1936–
Bio-Babel: can we survive the new biology?
Bibliography: p.
1. Bioethics. 2. Biology–Social aspects.
I. Title.
QH332.U97 174′.2 77–79595
ISBN 0–8042–0786–0

To my wife, Jan,
My children,
My mother.
May the smiles always win out over the tears.

In the Year 2525

In the year 2525, if man is still alive, if woman can survive, they may find—
In the year 3535, ain't gonna need to tell the truth, tell no lies,
Everything you think, do, and say, is in the pill you took today.
In the year 4545, ain't gonna need your teeth, won't need your eyes.
You won't find a thing to chew, nobody's gonna look at you.
In the year 5555, your arms are hangin' limp at your side,
Your leg's got nothing to do—some machine's doin' that for you.
In the year 6565, ain't gonna need no husbands, won't need no wives,
You'll pick your sons, pick your daughters too, from the bottom of a long glass tube.
In the year 7510, if God's comin', he ought to make it by then,
Maybe he'll look around and say, "Guess it's time for the Judgment Day."
In the year 8510, God is going to shake his mind and then,
He'll either say "I'm pleased where man has been," or tear it down and start again.
In the year 9595, I'm kind of wonderin' if man is going to be alive,
He's taken everything this old earth can give, and he ain't put back nothin'.

Now it's been 10,000 years; man has cried a billion tears. For what, he
Never knew; now man's reign on earth is through. But through eternal night, the twinkling starlight; so very far away, maybe it's only yesterday.

In the year 2525. . . .

—IN THE YEAR 2525
Words and Music by Rick Evans,

PREFACE

> Pleasure is a shadow, wealth a vanity, and power a pageant; but knowledge is ecstatic in enjoyment, perennial in fame, unlimited in space, and indefinite in duration. In the performance of its sacred offices, it fears no danger, spares no expense, looks in the volcanoe, dives into the ocean, perforates the earth, wings its flight into the skies, explores sea and land, contemplates the distant, examines the minute, comprehends the great, ascends to the sublime—no place too remote for its grasp, no height too exalted for its reach.
>
> —DeWitt Clinton

Knowledge has quadrupled in the last 25 years. As many as 90 percent of all the scientists who have ever lived in history are alive today. About 95 percent of all of the technological progress made in history has been made in the last generation. We have done all of the things which Clinton described. We have built a tower of knowledge! And thus, if Clinton was right in inferring that there is a direct relationship between knowledge, happiness, and contentment, our tower should have made us four times happier and more content than we were in 1950. But are we? We should be "ecstatic in enjoyment" and should have ascended "to the sublime." Have we?

Some people are happier and more content, of course, but, overall, every negative social indicator (crime, divorce, illiteracy, etc.) has increased in the last 25 years, often far out-stripping population growth. We seem to be in a period in which our morality, ethics, spirit, sense of purpose, etc., are being eroded rather than built up. All too often today we seem to be increasingly confused, discouraged, frustrated, and depressed. Instead of being ecstatically happy, all too often we simply seem to be psychologically numb, for our supreme adventures are being drowned out by a deluge of supreme problems. As Angelo J. Cerchione said in 1974, in a book titled *The Energy Crisis,*

> Energy Crisis! The bell rings and we stagger to the center of the ring, touch gloves . . . again and have at it.
>
> Bite down on the mouthpiece to ease the pain. Keep moving . . . keep swinging . . . what round is this, anyway? When did this fight start?
>
> It seems that we first came out of our corner in the '50s against a savage club fighter called discrimination. But there's been so many rounds since then . . . civil rights . . . peace . . . environment . . . women . . .
>
> See the drawn faces on television saying the same things they said at the beginning of all those other rounds. Hear the tense voices of the commentators raising the same old questions. Hasn't this all happened before?

It's becoming painfully obvious that we now live in the most abnormal period of time in the history of man. John F. Kennedy warned us of that fact when in 1963 he said, "We happen to live in the most dangerous times in the history of the human race. We are destined to live out most, if not all of our lives in uncertainty and challenge, and peril."

Our age is so dangerous because of the complex, technical, unpredictable, and global nature of our problems. In fact, our problems are now so big that there are no longer any obvious, easy solutions to them. We are living in the most muddled, confusing age in history! One can't help but wonder whether we may have constructed our tower of knowledge too hastily. Did we use enough wisdom in its construction—the mortar needed to hold it together? Have we actually produced another tower of Babel?

Unfortunately, although our tower is already quite shaky, the building goes on. In fact, the heaviest construction is yet to come! The tower may have to support the greatest scientific revolution of all time—the biological revolution—as it is added piece by piece in the next few decades. Will the tower hold? Or, could it turn into a Bio-Babel which could come crashing down around our ears? The purpose of this book will be to examine the number, nature, and "weight" of the biological "pieces" yet to be added, the current "load limit" of our tower, and ways in which it might possibly be "shored up" before those pieces are added.

I would welcome comments and correspondence from readers about this book or the biological revolution in general.

—A.R.U.

TABLE OF CONTENTS

PART I

THE NATURE OF THE BIOLOGICAL REVOLUTION

The really revolutionary revolution is to be achieved not in the external world, but in the souls and flesh of human beings.

—Aldous Huxley

INTRODUCTION

My greatest inspiration is a challenge to attempt the impossible.
—Albert A. Michelson

Every noble work is at first impossible.
—Carlyle

The world advances by impossibilities achieved.
—Emerson

About the year 1500, a seeress or prophetess by the name of Martha ("Mother") Shipton supposedly made the following predictions:

Carriages without horses shall go,
  and accidents fill the world with woe.
Around the world thoughts shall fly,
  In the twinkling of an eye.
Under water men shall walk,
  Shall ride, shall sleep, shall talk.
In the air men shall be seen,
  In white, in black, in green.

It's doubtful whether many people in Martha Shipton's time, or even for several centuries to come, ever took her words seriously. Most of those who heard about her predictions probably just laughed and dismissed the whole thing as "impossibility." But ridicule, or use of the word "impossible," usually results from intellectual provincialism. Those who rejected her prophesy lost sight of the fact that the 16th century was not an end point in history. Other centuries were ahead. By the early 20th century her predictions had all come true!

But even Martha Shipton's amazing vision only went so far. There were many consequences in her predictions that even she could not foresee. For example, could she possibly have foreseen the vast highway systems that her "carriages without horses" would bring into being? She did predict that "accidents would fill the world with woe." But could she possibly have foreseen the day when the "accidents" she described would have taken over a million lives around the world?

In 1865 another visionary, Jules Verne, wrote two amazing fictional books on space travel. The books were titled *From the Earth to the Moon* and *Voyage to the Moon.* In his "stories," Verne described three astronauts in a bullet-shaped projectile, shot out of a cannon from Stone's Hill, Florida. His three astronauts experienced weightlessness during their 97 hour and 10 minute trip to the moon. After planting the United States flag, they returned to earth, with "splashdown" and recovery operations taking place in the Pacific.

In 1865, and for many years thereafter, most of those who read Verne's "stories" probably considered them simply entertaining fiction. Although intrigued by the idea of space travel, most people undoubtedly considered the whole thing to be basically "impossible." In fact, even as late as 1957, one month before Sputnik, one of England's leading astronomers, Dr. Richard Wooley, publicly declared the idea of space travel to be "utter bilge." But in 1969 travel to the moon suddenly became reality and Jules Verne turned out to be more of a prophet than a storyteller. For example, Stone's Hill, Florida is only about 50 miles from Cape Kennedy, where the three astronauts of the Apollo X mission were shot into space in 1969. The Apollo X flight took 98 hours and 20 minutes, during which time the astronauts did experience weightlessness! The United States flag was planted on the moon, and "splashdown" and recovery did take place in the Pacific!

But despite his uncanny vision into the future, Jules Verne also vastly underestimated that future, just as did Martha Shipton. For he did not foresee the new propulsion systems, new materials, new medical techniques, new knowledge, etc., that would result from going to the moon. Nor did he foresee the billions of dollars involved, subsequent planned, manned trips to Mars and beyond, the possibility that moon bacterial organisms might contaminate the earth, etc.

Radioactivity was discovered about 1900 by Henri Becquerel, the Curies, and others. In the October 4, 1903 issue of the *St. Louis Post-Dispatch,* an editorial asked whether the newly-found radioactivity might not be the source of an inconceivable new power that could be used in warfare and as a possible instrument for the destruction of the world. Most of those people reading the editorial probably found it interesting, but then, undoubtedly, dismissed it as science fiction and "impossibility." In fact, as late as 1937, even Nobel physicist and nuclear pioneer Lord Ernest Rutherford publicly declared his belief that the energy associated with radioactivity would never be harnessed. However, in 1942, it was done below Stagg Field in Chicago, and in 1945 "the ultimate weapon" was dropped. The editorial had been right! But, even so, the editor involved had also vastly underestimated the future. For although he had mentioned "the destruction of the world," it's doubtful whether he actually could have conceived of the awesome power of the atomic bomb and the reality of the horrors it produced at Hiroshima and Nagasaki. Certainly the hydrogen bomb, nuclear testing and the fallout problem, the proliferation of nuclear weapons and the nuclear arms race, and the peacetime uses of radioactivity and nuclear power were beyond his prophetic ability.

The three historical excerpts just cited were selected with a specific purpose in mind. These excerpts, and the many others that could also be related, point out four extremely important facts that should be clearly recognized about scientific progress by everyone reading this book. If these facts are not recognized, the book may not seem to describe potential reality at times and may not have the overall impact that is intended by the author.

1. NO MATTER HOW "IMPOSSIBLE" SOMETHING SOUNDS, IF IT CAN BE MENTALLY VISUALIZED OR PICTURED, IT SHOULD PROBABLY BE PLACED IN THE REALM OF POSSIBILITY.

It's becoming increasingly clear that those who use the word "impossible" are more and more being proven wrong later. In fact, today, even while one man may be labeling something "impossible," another may be in the process of doing it.

2. ONCE MANKIND REALIZES SOMETHING CAN BE DONE, SOONER OR LATER IT PROBABLY WILL BE DONE.

The pace of scientific discovery and the accumulation of knowledge is quickening (knowledge is currently doubling about every five to seven years). There's a great deal of validity in the boast of some that given enough time, money, and manpower, science could solve any problem. However, because of the increasing tendency toward unexpected "sudden breakthroughs" and the "threshhold effect," the widening scope and thus increasing impact of scientific discoveries, an accelerating rate of change which may be causing increasing "future shock," a seemingly increasing tendency toward a "live for today" philosophy, the increasingly complex communication gap between scientists and laymen, and man's continued, apparently instinctual reluctance to change and to face new social responsibilities, current scientific discoveries seem to be reaching a society increasingly unprepared for them.

3. THE TIME SPAN BETWEEN IMAGINATION AND SCIENTIFIC REALITY IS CLOSING.

First of all, the imagination-discovery gap is closing at an accelerating rate. The gap was 400 years for Martha Shipton's predictions, 104 years for those of Jules Verne, and 39 years for the predictions in the *St. Louis Post-Dispatch.* Secondly, the discovery-application gap is closing at an accelerating rate. For example, for photography it was 112 years, the electric motor 65 years, the telephone 56 years, radio 35 years, 18 years for the x-ray tube, 12 years for television, 10 years for nuclear reactors, 6 years for the atomic bomb, 3 years for most plastics, etc. Overall, yesterday's science fiction and "impossibility" are becoming today's scientific reality at an accelerating rate.

4. WHEN MAN ATTEMPTS TO ESTIMATE THE FUTURE, HE ALWAYS UNDERESTIMATES IT.

Many major discoveries are made almost by accident and come about with little or no advance warning. For example, x-rays, nuclear energy, radio, television, quantum mechanics, relativity, transistors, lasers, atomic clocks, electronics, penicillin, the Van Allen belts, etc., were basically unexpected discoveries that were largely unforeseen. But even when society and/or the scientific community is fully aware of the probable advent of a certain discovery (a rarity!), the full consequences of that discovery are never foreseen. Some of the reasons for this lack of foresight have already been listed above. Overall, it appears that most people (including most scientists) are simply poor prophets!

The scientific discoveries predicted by Martha Shipton, Jules Verne and the *St. Louis Post-Dispatch* brought unexpected, earthshattering consequences when they were made. Man just wasn't prepared for the reality of these "impossibilities." In the late 1970's, man finds himself within the imagination-reality gap of another scientific revolution of possibly even greater impact—the biological revolution. As the noted scientist, Arthur Koestler, said several years ago: "Biology is just reaching the critical point of sudden acceleration which physics reached a generation ago. . . . The biological time bomb is about to explode in our face."

If Koestler is right, the explosion he predicts will probably once again reach a largely unprepared society, for most of society today seems to be either completely or at best only partially aware of the nature, scope, and potential meaning of the biological research now in progress. There are a small number of scientists and laymen who have issued predictions and warnings about the potential societal problems that may result from such work. However, others have expressed their belief that it is "unlikely" or even "impossible" that certain aspects of the research will ever come to fruition. Controversy and divergence of opinion always nullify concern. The warnings have been largely neutralized!

Overall, ignorance, disinterest, and apathy seem to be outweighing concern and preparation for the biological revolution, just as they have with so many other scientific revolutions in the past. In the 1970's, humanity still apparently hasn't fully recognized the reality of the four points cited previously, and still seems to suffer in large measure from intellectual provincialism.

And so, once again, another scientific revolution could be "sneaking up on us" and we could once again be suddenly confronted with extremely serious (and often unwanted) questions, issues, and crises for which we are not ready. As noted biochemist James W. Cobert, Jr., of Michigan State University, said recently: "The biological age is upon us and what is to happen in the next 30 years will produce a cultural shock no one in the Western world is prepared for."

Can we afford to gamble that Koestler, Cobert, and others are wrong? Even if they were only half-right, or less, it would seem that as many people as possible must be informed about the biological revolution—as soon as possible! Unfortunately, there are so many, often interrelated, facets to the subject that even trying to categorize or outline the biological revolution can result in confusion. In Part I, the biological revolution has been arbitrarily (and also rather uniquely) divided into five parts. Each of the five chapters in Part I discusses one of the five parts.

CHAPTER 1

REPRODUCTION

And here's the happy bounding flea—
You cannot tell the he from she.
The sexes look alike you see;
But she can tell, and so can he.
—Roland Young, *The Flea*

For thousands, perhaps even millions, of years, the sex drive and the urge to reproduce have strongly influenced and/or controlled the minds and bodies of much or even most of mankind. In the 1950's and 1960's, the tide began to turn, however. We are now learning how to chemically *control* our reproductive processes and even our sex drive. That control could be largely complete by the year 2000, as outlined below.

FERTILITY (BIRTH) CONTROL

Human reproduction is a complex, delicate, and exquisitely balanced process in which the central, determining event is the union of a male cell (sperm) with a female cell (egg) to form a primary cell (fertilized ovum). The subsequent division and duplication of that primary cell, up to approximately 1×10^{14} times (100 trillion), results in the formation of an embryo, fetus, baby (about 2 trillion cells), and finally adult human being (60–100 trillion cells).

Before 1960, attempts to prevent human reproduction centered around trying to prevent the union of egg and sperm with such mechanical devices as condoms, diaphragms, uterine rings, etc. However, as early as the 1920's and 1930's, medical researchers had learned that reproductive ability is governed by the flow and interaction of hormones and other chemicals that regulate the production of sperm in the male and the release of an egg into a receptive uterine environment in the female. Specifically it was found that the menstrual cycle is largely controlled by two hormones, estrogen and progesterone, which are produced mainly in the ovaries. Today, we know that it is the interaction of these two hormones with two others, known as FSH and LH, which are produced by the pituitary gland, that causes the menstrual cycle, that causes ovulation, and that enables the reproductive system to foster and encourage conception and implantation of the fertilized ovum on the uterine wall.

Estrogen is released in the first half of the cycle and serves to prepare the uterus surface to receive the egg. Progesterone is involved in the release of the eggs from the ovaries. If the egg is not fertilized at the end of two weeks, progesterone production stops and menstruation begins. Menstruation is simply a discharge of the womb coat which had been prepared for the egg.

It became obvious to researchers even before 1940 that doses of estrogen and progesterone from outside the body might stabilize the menstrual cycle, prevent the release of eggs from the ovaries, and thus prevent conception. In essence, the increased supplies of estrogen and progesterone would "tell" the body that pregnancy had occurred. In the 1940's and 1950's, the many technical difficulties involved in preparing an oral, synthetic estrogen-progesterone birth control pill were worked out and the "pill" was clinically tested. It reached the general public in 1960, and has since been used by over 10 million women in the U.S., and more than 50 million in the world.

Pregnancy, although a natural condition, is known to be associated with certain health hazards. Possible complications and side effects include severe vascular headaches, attacks of migraine, worsening of epilepsy, changes in the blood and its vessels, nausea, and vomiting. The pill simulates pregnancy and thus suppresses the release of an egg from the ovaries. A woman taking the pill partially mimics the entire biological course of pregnancy as many as eleven times a month, rather than at most once in 320 days. It was, therefore, suspected almost immediately that there might be possible side effects associated with usage of the pill.

The first definite information of this nature came in the late 1960's when British researchers documented an increased incidence of blood-clotting or thrombosis among pill users. After hearings on the subject in 1970, the U.S. Congress declared that the benefits of the pill outweigh the risks. The pill was declared legally safe, although it was suggested that warnings be attached by the manufacturers. (Note: Among a million women 20 to 34 years old, the British found that the number of deaths from abnormal blood-clots was two for women not using the pill and 15 among those who did use it. There were 228 deaths related to pregnancy.)

In the 1970's, researchers are attempting to document further suspected side effects. Many believe there may be additional links between the pill and hormone imbalance, atherosclerosis, gall bladder disorders, vaginal infections, liver tumors, neurological disease, strokes, diabetes, high blood pressure, fluid retention, nausea, headaches, and depression. In the fall of 1975, the Federal Drug Administration felt enough evidence had accumulated to draft new warning labels for birth control pills, to alert physicians and patients to some new possible adverse effects in addition to blood-clotting. The new labels include warnings to the effect that: (1) women over the age of 40 should use other contraceptive methods, if possible, to avoid an increased risk of heart attacks; (2) although still extremely rare, users may face an increased risk of nonmalignant liver tumors; and (3) users should wait at least three months before attempting conception after discontinuing use, in order to avoid an increased risk of spontaneous abortion.

In February of 1976, three drug manufacturers withdrew their sequential birth control pills from the market. The sequential birth control pill delivers estrogen alone for the greater part of the cycle in which it is taken, and progesterone alone for the remainder. Food and Drug Administration officials said the sequential pills carried a higher potential for cancer of the uterus, and a higher risk of pregnancy and blood-clotting than combination pills. The officials estimated that 5 to 10 percent of the approximately 10 million women using birth control pills in early 1976 were taking sequentials.

In April of 1977, the Population Council, a private research organization, announced the discovery of a synergistic effect between the pill and smoking in women over 40. According to their recent study, women who both smoke and take the pill suffer a sixfold increase in the death rate from heart attacks and strokes over those who only smoke.

The known and suspected side effects of the "pill," coupled with the self-discipline needed in taking daily oral doses (the pill is not for the absent-minded!), have made it increasingly clear to the general public that it is far from being an ideal contraceptive. This realization is undoubtedly why sales have stabilized in the 1970's, despite the fact that the number of women in the 15 to 44 age range has increased several million. Confidence in the pill is also apparently dropping in the medical community, for many researchers are already working on second and third generation contraceptives which can hopefully be used to replace it.

For example, an oral estrogen-progesterone "monthly pill," a similar estrogen-only "pill," and several variations of a progesterone-only "mini-pill" are currently under development or in the testing stage. And, in late 1975, the Alsa Corporation of England marketed a membrane-enclosed reservoir of progesterone which can be placed directly in the uterus and supposedly provide protection for up to a year.[1]

It appears, however, that birth control injections that need to be administered only once every three months to a year, or more, may ultimately hold more attraction as practical birth control agents. In fact, such injections are already available in such countries as Mexico, the Netherlands, West Germany, and Thailand. In the United States the first such injectable drug was approved by the FDA in late 1973 for limited use in the treatment of uterine cancer. The drug, know as Depo Provera or DMPA, is made by the Upjohn Company, which claims it is also an excellent contraceptive that only has to be injected once every three months.[2] It is possible that DMPA could be cleared by the FDA for commercial production as early as sometime in 1977.

Several techniques are also in the testing stage for injecting a single capsule or cluster of progesterone capsules under the skin (in the lower back or elsewhere) with a hypodermic needle, in a doctor's office procedure. Such injections could produce anything from an "annual contraceptive" to one effective for up to six years or more. Unfortunately, however, they appear to be at least several years away from commercial production. Although the short-term side effects of progesterone-only drugs appear to be less than those of the current pill, their possible long-range effects still remain largely unknown.

Other researchers are working on yet other types of immunization. They are developing injections of antibodies that either coat an egg so that it cannot be penetrated by sperm, or interfere with human chorionic gonodotropin (HCG), a hormone essential in early pregnancy. Many believe that such an "anti-pregnancy vaccine" can be developed that would not interfere with ovulation, menstruation, or other bodily functions and would be effective for a year or more. They also believe it could be done within five years with a concerted effort.[3]

In 1971, Dr. Andrew Schally and co-workers announced that they had determined the structure of and had synthesized a hormone known as LH-RH/FSH-RH.[4] This hormone controls the synthesis and release of two pituitary hormones, LH and FSH, which in turn control menstruation, ovulation, and the overall operation of the reproductive system in women. Schally and others believe a "peptide pill" or immunization may eventually be found which effects birth control for months at a time by interfering with LH-RH/FSH-RH, and thus indirectly with LH and FSH.

There are also several non-hormone chemicals under investigation as birth control agents. The most promising belong to a class of chemicals known as the prostaglandins (PG). These chemicals are synthesized by various glands in the body, mainly the prostrate gland in the male, and thus their name. Six of the twenty prostaglandins that had been discovered by 1975 appear to be more important in the human body. Outside doses of one of the six, PGF_2 alpha, can apparently cause restriction of the outflow of blood and hormones from the ovaries and can thus induce menstruation, even if a fertilized egg is present. PGF_2 alpha can thus be used to prevent ovulation and may have fewer side effects than hormone contraceptives. Of course, PGF_2 alpha can actually be considered to be a "morning after" contraceptive, and it has been ballyhooed as such in the popular press.

The development of male chemical contraceptives has been plagued to date by such physical side effects as loss of desire, loss of potency, and an increased sensitivity to alcohol. Interestingly enough, a man could take his wife's pill and lose his fertility within two weeks. But he would also lose his libido! Such side effects have thus produced a low psychological receptivity among men for male contraceptives. Some have said, at least semi-seriously, that an effective male contraceptive will never be produced as long as virtually all of the research needed is done by men. However, others have added, again only partly tongue-in-cheek, that if a male birth control agent could be developed that also increased libido, it might be a different story!

Unfortunately, in addition to the physiological and psychological problems involved in developing a male contraceptive, there are also some serious research obstacles. For example, there are more places in which one can interfere with the female reproductive system than in the male—in large measure because scientists know considerably more about the female reproductive system. At the present time, researchers are limited to attempting to interfere with sperm production in the male. This, of course, also limits them to examining masturbation samples of sperm in order to determine the effectiveness of the contraceptive being tested—a less than ideal testing technique. There's also the very basic problem of even finding male research subjects for fertility research. Adequate numbers of female volunteers can usually be found through gynecologists, obstetricians, Planned Parenthood Associations, etc. But male volunteers can frequently be found only in prisons or in the army.

Nevertheless, despite the many problems involved in developing a male contraceptive, there is progress being made. In fact, there are actually a number of male "pills" at various stages of development, including clinical trials. The most promising may be a combination of a synthetic derivative of estrogen and a synthetic derivative of the male hormone, testosterone. Problems with this agent currently include expense, effectiveness, and high blood pressure, but most expect them to be resolved within a few years.[5] Other agents under investigation include an analogue of the common sugar, glucose,[6] a series of synthetic steroids,[7] and IH-imdazole-3-carboxylic acids and their derivatives.[8] There are also researchers attempting to turn off sperm production with heat, infrared radiation, and ultrasound.[9] With work pregressing on several fronts at the present time, it would appear that the commercial availability of a male contraceptive may be only a few years away.

The development and improvement of mechanical contraceptives is also continuing. In recent years, the most extensive research in this respect has centered on the intrauterine device (IUD). Medical historians believe the first

IUD was probably simply a crude stone used in ancient times to prevent pregnancy in camels. In the twentieth century, IUD's made of plastic and metal have become available for use—in a wide variety of shapes and sizes. Over nine million IUD's have been used since 1966 in the United States, with over four million currently in use. Bleeding, cramps, pain, and expulsion from the uterus have frequently been associated with the use of IUD's such as rings and loops. However, in 1974, IUD's were also linked by some to uterine perforations, serious infections, blood poisoning, miscarriages, and even death. The same sort of statistical controversy over safety arose with IUD's that is still waging over the pill. That controversy may not be resolved for some time.

In the meantime, since the controversy arose, several new variations of IUD's have been marketed. One has a copper wire wrapped around the stem. The copper slowly leaches into the wearer's uterus, supposedly improving the contraceptive properties of the IUD.[10] Another recently commercialized IUD slowly released progesterone. Overall, however, it remains to be seen whether the most recent IUD's are actually safer and improvements over earlier models.[11]

In summary, many more contraceptive chemicals and techniques are under study in addition to those already discussed. However, although fertility control research is currently interwoven with bureaucractic and public hypertension, and the money available for such research declined in 1975 and 1976, it seems reasonable to predict that the pill of today is a mere forerunner of more sophisticated agents yet to come. The development and commercialization of such agents should be largely completed by 1985, making the regulation of fertility much easier, surer, and more precise than it is currently.

ABORTION

So much has been written elsewhere about the means of abortion available today (mainly surgical or mechanical) and the moral and ethical questions involved in using such means, that nothing further in this respect will be added here. However, a comment or two about the possible development of *chemical* abortifacients should be made.

Drugs such as Oil of Wintergreen have been used illegally for many years in attempts to produce abortion. Unfortunately, however, such attempts have frequently ended in death. The search for a legally-acceptable chemical abortifacient was opened in the 1950's and 1960's in a rather unique way, with a rather unique drug known as diethylstilbesterol (DES). When fed to cattle in small quantities, DES was found to have the interesting and also economically-valuable property of enhancing their rate of growth. The drug was ultimately fed to as many as three-fourths of all of the cattle slaughtered in the United States in the late 1960's. However, DES was soon subsequently found to be carcinogenic to humans, and its presence in meat (although not its use) was forbidden by the Federal Drug Administration. Enforcement proved difficult, however, and the use of DES was banned in 1972, after several countries banned beef imports from the United States.

Not only was DES found to function as a growth hormone in cattle in the 1960's, but it was also found to perform like a female sex hormone in humans. In *small* doses it was found to be an effective postcoital contraceptive—a so-called "morning after" pill—if taken within three days after sexual relations. It was used rather extensively in this respect for several years before the Federal Drug Administration approved its use in 1972 in "emergency cases" of rape or severe mental anguish. Interestingly enough, DES was also used in the 1960's in *large* doses, with a fairly limited number of patients, to prevent miscarriages during early pregnancies. However, in the 1970's, it was discovered that over 50 of the daughters born of those pregnancies suffered from a rare form of vaginal cancer in their teenage and early adulthood periods. The link between cancer and DES usage in both cattle and humans has now clearly been established.

Technically, of course, DES is not a true abortifacient, for it does not dislodge an already implanted fertilized egg. It rather prevents implantation of that egg. However, the difference is fairly subtle since it has been labeled the "morning after" pill. And thus, although the future doesn't look too bright for DES at the moment, it has "opened the door" to the development of other, true abortifacients.

Current research efforts to find such agents are directed two ways; toward finding agents that can interfere with embryonic development and toward finding agents that can cause expulsion of the embryo or fetus. Research in the area of embryonic development is still in its infancy, and no agents have as yet been found which do not cause serious teratological side-effects (fetal malformations). Unfortunately, an embryonic development agent must probably have an abortive efficiency in excess of 99 percent, before teratological side effects could be largely overlooked. This requirement makes even human clinical testing very difficult, for the experiments performed have to be backed up by the possibility of physical abortion.

The likelihood of success in developing an agent that actually causes expulsion of an embryo or fetus, rather than interfering with its development, appears to be considerably more promising. In fact, two such agents, the prostaglandins PGF_2 alpha and PGE_2, are already on the market in several countries. Both prostaglandins are commercially available in Great Britain for inducing labor and terminating pregnancy in the second trimester. The Japanese have marketed PGF_2 alpha. In the United States, PGF_2 alpha is available on a restricted basis for use in second trimester abortions.[12] It should be pointed out, however, that there is still work to be done on both prostaglandins with regard to teratogenesis and other possible side effects.

Other types of chemical abortifacients may also be on the way. For example, Dr. Roger Short and other researchers at the United Kingdom Medical Research Council's reproductive unit at Edinburgh reported in May of 1975 that they had developed an injection (chorionic gonadotropin antibody) which inhibits the release of progesterone, which is essential for embryo development.[13] A single injection in monkeys has terminated an early pregnancy, and also prevented others for at least a year. Work is continuing on the agent, although a commercial antipregnancy vaccine and abortifacient is apparently still several years off.

Extensive progress in the development of chemical abortifacients can probably be expected in the next decade or so, although social pressure, both for and against the use of such agents will, undoubtedly, effect the developmental times involved.

CONTROL OF SEXUAL DESIRE

Despite many myths to the contrary, there are currently no chemical aphrodisiacs known which combine reasonable safety and effectiveness. Alcohol, marijuana, LSD, methaqualone (Quaaludes), etc., have been called "love drugs" by some, but most scientists believe they actually lower inhibitions rather than stimulate sexual desire. However, even if they do increase desire, they certainly couldn't be classified as being safe, effective, or acceptable aphrodisiacs.

Accelerating research into human reproduction will probably bring, indirectly, an increased understanding of the physiology involved in sexual desire. The brain centers which control and mediate libido have already been identified. With more specific information available about the chemistry and chemicals involved, it seems safe to predict that specific drugs selectively regulating sexual desire will be developed. Regulating drugs of this type may be at least one or two decades away, but the situation could change suddenly with an unexpected breakthrough. For example, a chemical called L-Dopa was announced as being found effective against Parkinson's disease in 1969. The development of hyper-sexuality was listed as being one of its side effects, although no further research in this respect has been reported.

ARTIFICIAL INSEMINATION

An artificial insemination donor process (AID) has been used for over 15 years in the United States to produce children in marriages where the husband is infertile. In the process, as it currently operates, the doctor alone selects the sperm donor, who remains anonymous. The sperm is used almost immediately after collection, or it is stored for a short period of time before use. Estimates of from 100,000 to over 250,000 have been given for the total number of children believed to have already been conceived in the United States by means of AID, with over 10,000 more such conceptions now occurring annually.

The biggest technical problem involved in AID is simply the necessity of having to use the donor sperm almost immediately. If long-term sperm storage were possible, AID could obviously become more flexible, for close proximity in space and time of donor, recipient, and middleman (doctor) would no longer be required. Increased flexibility could also lead to greater selectivity, for both the parents and the doctor.

The idea of storing frozen human sperm is not new. It has been around since the 1950's, when techniques were developed that allowed the storage of frozen bull semen for over ten years, without genetic damage or more than slight loss of vitality. The extent and selectivity of these techniques have increased to the point where over 5 million calves are now conceived each year at "breeding farms" from frozen bull semen. Human sperm, however, is far more fragile than that of cattle. That fact, combined with an apparent lack of enthusiasm for abolishing the need for intercourse for conception, greatly discouraged and limited similar research on human sperm for many years.

Predictions about the establishment of human "sperm banks" remained largely unfulfilled until the late 1960's and early 1970's. But then, driven by a complex, interrelated set of motivational factors, many Americans decided, in almost a quantum jump, to adopt alternative styles of life other than marriage, and to limit and space the number of children they had when they did get married. Concurrent with this development, worries about the safety and reliability of various birth control methods began to appear, the relative inexpensiveness and ease of sterilization became increasingly apparent, and concern and even fear about bringing children into the world began to increase. One of the major results of the convergence of these and other factors was a rather dramatic increase in the number of vasectomies performed. According to various estimates, there were upwards to a million vasectomies performed in 1970, versus about 40,000 in 1960! Some estimates place the total number of vasectomies performed in the 1970's through 1976 at more than five million.

The trend toward more vasectomies has been at least partially responsible for another trend in recent years. Realizing that a vasectomy is essentially irreversible, an increasing number of men have been taking out "fatherhood insurance" before the operation. The "insurance" involves storing some of their frozen sperm, apparently in case they later change their minds. As the word has spread about this option, an increasing number of men not having vasectomies have also been storing some of their frozen sperm for a variety of other reasons. Some of the reasons are quite justifiable, like trying to concentrate the sperm of men with digospermia (low sperm count). Others are quite unrealistic, like the belief of some men that sperm banked early in their lives will be superior to that which they will produce when older. But, for whatever the reasons, the storing of frozen sperm has increased dramatically in the 1970's.

The trend toward increasing vasectomies is easier to understand than that of increasingly storing frozen human sperm. Although some have claimed that human sperm can now be frozen and stored for over ten years, thawed, and then used to successfully impregnate a woman, other estimates have been as low as 16 months. In February of 1972, the American Public Health Association's Council on Population publicly expressed reservations about the effectiveness of sperm frozen more than 16 months. Estimates of the number of births that have resulted from the use of frozen sperm also vary. Estimates ranging anywhere from 100 to over 400 have been given.

Nevertheless, it seems quite likely that the indefinite storage of frozen human sperm may be right around the corner. One indication this may be true may be the great strides that have been made in recent years in storing frozen blood. Both red blood cells and sperm have been frozen in glycerol for some time now to prevent freezing damage from the formation of ice crystals. However, recent improvements in the glycerol technique now allow blood storage for over ten years, with some claiming indefinite storage to be possible. It seems likely that at least some of the same techniques will be applicable to the storage of sperm.

Once long-term "sperm banks" become a reality, it seems reasonable to expect that artificial insemination will be placed on an increasingly selective basis, just as it has been with cattle. Instead of anonymous donors, a trend will undoubtedly develop to keep complete dossiers on donors and then to select the donors with certain qualities and characteristics in mind. Potential parents could thus be given a choice as to the physical qualities of their offspring; the temptation would then be almost overwhelming to attempt to do the same thing with intellectual and personality characteristics. If a semen donor were necessary, what parent wouldn't want the best possible or most desirable children, if he knew he had a choice? There might even be a trend develop to increasingly select famous and/or admired people as donors. As the degree of possible selectivity increased, the number of parents who chose to use AID, even though they didn't actually need it, might also increase.

CHOICE OF SEX IN OFFSPRING

The desire of parents first to predict and then to control the sex of their children has been one of the most enduring of human desires. In 1977, biologists began to routinely fulfill that age-old wish.

It is now known that the sex of a human (and many animals as well) is determined at conception at the chromosomal level. It is also known that the father rather than the mother controls the determination. Sperm can contain a female component (an X chromosome) or a male component (a Y chromosome). Sperm containing an X chromosome confer female sex on the eggs they fertilize, with male sex being conferred by sperm containing a Y chromosome. Thus, if a fetus contains two X chromosomes in each of its cells, it will develop into a female. If the fetus contains one X and one Y chromosome, it will develop into a male.

By the early 1970's, promising sperm separation results had been obtained with rabbit, rat, and cattle sperm, with electrophoresis, centrifuging, and chemicals. In 1972, A.M. Roberts of Guy's Hospital Medical School in London discovered that human sperm bearing X and Y chromosomes can be distinguished by their swimming abilities.[14] "Sleeker" Y sperm swim faster than X sperm, apparently because X sperm contain roughly 3 percent more genetic material, and are thus heavier. It is probably the faster swimming male sperm that result in male babies outnumbering female babies in the ratio 106 to 100.

In November of 1973, biologist Ronald J. Erickson and co-workers at the A.G. Shering Company in Berlin announced the discovery of a technique for progressively separating human X and Y sperm through repeated placement of the sperm mixture in a dense solution that resists the swimming of the X sperm more than that of the more mobile Y variety.[15] The German researchers claimed to have been able to produce healthy, fertile sperm that were up to 85 percent the Y variety. Some researchers subsequently challenged those claims, and termed them "unsubstantiated." However, Erickson is currently associated with Giametrics Limited of California, where he is using the swimming technique to help the estimated 20 percent or so of all couples who are unsuccessful at reproduction attempts. Because there is supposedly about a 90 percent chance of a boy baby resulting from the procedure, he and several other clinicians in the United States and Europe are also currently using it with the goal of producing male children! Erickson plans to broadly market the technique in 1978.[16] It will also undoubtedly be increasingly used in livestock insemination in the late 1970's.

There may also be another dramatic key to the whole problem of sex determination soon available. In December of 1973, the same German researchers just mentioned reported that strong antibodies directed against certain substances on the surface of the Y sperm of mice reduced the likelihood of their successfully fertilizing an egg and hence producing male offspring. (See footnote 16.) This work essentially now takes the whole matter of sex determination a step beyond the chromosomal stage, to the gene stage, for the work indicates that there may be a major sex determining gene in Y chromosomes. In November of 1975, Stephen S. Wachtel and several of his colleagues at the Memorial Sloane-Kettering Cancer Center confirmed the existence of that gene.[17] Wachtel and his co-workers have speculated that it may determine the development of undifferentiated sex organs into either male or female organs (e.g., testes rather than ovaries). In October of 1976, the same research team announced that they had made important advances in locating and perhaps eventually isolating the male sex gene.[18]

The gene that Wachtel and others have discovered is the first sex gene to be assigned to the human Y chromosome. Park B. Gerald, a physician at the Children's Hospital Medical Center in Boston, has called the discovery "a major event in human genetics." (See footnote 17.) The full significance of the discovery, including its possible role in sex determination, is not easy to predict at the moment, however.

It is difficult to believe that a technique of human sex determination could be developed and in clinical use by mid-1978, when the underlying basis for the technique was only discovered in 1972. It is now quite clear, however, that choice of sex in offspring research is no longer in its infancy, but has come of age and is maturing fast!

ARTIFICIAL INOVULATION

It is only one step from artificial insemination (the fertilization of an egg inside the mother's body) to artificial inovulation (the fertilization of an egg outside the mother's body). It is only one more step to removing an egg from one mother, fertilizing it, and placing it in the body of a foster mother for development and birth. Experiments such as these with animals date back to the early 1960's.

In one of the earliest of such experiments, in 1962, two fertilized eggs were removed from two English sheep, and were then placed in the oviduct of a living rabbit. The rabbit was transported to South Africa, where the eggs were implanted in two South African ewes, which then gave birth to two English lambs.

In 1966, Dr. E. S. E. Hafez, an experimental biologist at Washington State University, commissioned a scientist friend from Germany to send him a hundred head of prize sheep. The entire herd was delivered to Hafez as several-day-old embryos placed inside the womb of a female rabbit! Upon delivery, each embryo was removed and planted in a ewe, where it gestated, and in a few months was born.

M.C. Chang of the Worcester Foundation and Benjamin Brackett at the University of Pennsylvania have been independently studying artificial inovulation in rabbits, mice, rats, hamsters, guinea pigs, and cats for almost 20 years.[19] The primary purpose of their research has been to find ways to prevent sperm penetration and thus possibly develop a new type of human contraceptive. However, a fourth or more of the inovulations have produced live offspring. Brackett also claims to have accomplished successful artificial inovulation into substitute mothers with rodents.

In April of 1973, International Cryo-Biological Services, Inc., located at

St. Paul, Minnesota, announced the discovery of a process they called "bovine ova transfer."[20] In the process, multiple-egg release is stimulated in "high-quality cows." The eggs are then artificially inseminated with sperm from "superior bulls." The fertilized eggs are removed, and each is transferred to a "less valuable incubator cow," which carries the embryo to birth. One "high quality cow" can thus supposedly produce up to 14 superior calves per year in contrast to a normal average of 3.5 during her whole lifetime!

Using the same technique, Home Farm in Sussex, England, owned by Genetics International, has publicly claimed the ability to now ship a "test tube of heifers" almost anywhere in the world. Upon arrival, the fertilized eggs, shipped in a nutrient fluid, are reimplanted in "incubator cows."

In May of 1973, a governmental agricultural laboratory in Cambridge, England, announced an interesting "twist" in producing "test tube cattle."[21] The twist involved freezing the fertilized cow eggs to −196° Centigrade (the temperature of liquid nitrogen) for six days before thawing them and reimplanting them into the uterus of a foster mother cow. The first artificial inovulation involving deep-frozen embryos and foster mothers has apparently already been accomplished in large mammals!

The artificial inovulation work with cattle just described is quite dramatic and has tremendous promise. However, it should also be pointed out that very little subsequent information has appeared since the initial announcements of several years ago. But, whether problems with consistency or even duplication of results have actually arisen is unclear at the moment. Some have claimed that the experiments have proven to be irreproducible. Whatever the trouble, if there is any, it seems likely to be cleared up with time.

Part of the foundation for such optimism can be found in the fact that artificial inovulation has been demonstrably successful with the frozen embryos of two small mammals. In January of 1975, R. R. Mauer of the National Institute of Environmental Health Sciences, and Harvey Bank of the Medical University of South Carolina announced the results of successful research with rabbits.[22] They reported freezing embryos at liquid nitrogen temperatures (−196° C.) from 30 minutes to two weeks. Upon thawing and culture growth, more than half supposedly continued to divide and grow for several hours. Implantation supposedly produced some viable fetuses and normal offspring. The results of similar research with mice will be discussed in chapter four.

> Laboratory lights glare down. Rubber-gloved hands glisten as they reach over the operating table and pluck out two ovaries, two oviducts and a uterus. The female organs are rushed to the Womb Room, a warm, humid chamber behind glass. There, eggs are flushed out of the organs and popped into a dish that contains sperm and chemicals. In three hours sperm penetrate the eggs. Within nine hours the eggs are fertilized, then divide into two cells, four cells, eight cells, sixteen. When they reach 16 cells, or the blastocyst stage (where the fertilized egg

is ready to implant itself in a uterus and become a fetus), the eggs are put into natural or substitute mothers who will carry them to birth. . . .

—*Science News,* Volume 103, February 24, 1973

The techniques described in the scenario above have already been used successfully with animals. With some modification, they should work just as well with humans. However, artificial inovulation research on humans has been understandably more cautious—particularly in the United States.

In the 1940's, Dr. John Rock of Harvard, the birth control pioneer, conducted some early experiments in which he obtained eggs from female cancer patients, artificially-inseminated them in a test tube, and was able to bring them to a three-cell stage. In the 1950's, Dr. Landrum Shettles of Columbia University established more precise conditions for test tube fertilization and grew several fertilized embryos, in a special culture, to the blastocyst (blastula) stage (16 cells and up).

In 1969, Drs. Patrick Steptoe, Robert Edwards, and Barry Bavister of Cambridge performed the first successful culture-dish fertilizations of human eggs in England.[23] They have performed several hundred since. They have also made about two hundred reimplantation attempts in the last several years, but have to date reported no successful pregnancies. Apparently, the fertilized egg, injected by means of a plastic tube into the uterus, is usually simply flushed away with the women's next menstrual period.

In 1971, Dr. Shettles reported that he had removed an egg from a woman undergoing an operation to correct a defect in her fallopian tubes.[24] The egg was fertilized with her husband's sperm, grown in a culture to the blastocyst stage, and was then implanted into the uterus *of a second woman.* Two days after the implant, a previously-scheduled hysterectomy was performed on the recipient. Examination revealed the egg had implanted properly and had grown to a size of several hundred cells.

In 1973, Professors Carl Wood and John Leetong, of Queen Victoria Hospital in Melbourne, Australia, reported they had performed experiments similar to those of Steptoe, Edwards, and Bavister, but with similar unsuccess, also.[25] In the early 1970's it was beginning to appear that artificial inovulation was stuck on "dead center." It could be routinely performed, but couldn't be brought to a culmination.

The agenda for the 1974 British Medical Association's annual meeting held in July in Hull seemed to indicate it would be a rather routine affair. But during a question period following a rather routine report on embryo implants, Dr. Douglas Bevis of Leeds University declared unexpectedly that three children, the world's first "test-tube babies," had been born in the previous eighteen months! Bevis said the births had been to women whose ova had been fertilized in a culture dish and then had been reimplanted into their wombs. However, Bevis then became mysteriously quiet about the announcement and very elusive about further details. He refused to identify the children, except to say that one was living in England and the other two in Europe, and that all three were perfectly normal. After three days of furor, Bevis admitted that he had personally participated in the research leading to the births, but he still refused to identify the other doctors involved, or the patients. Bevis has since announced that he would abandon further research of this type. He claims he has been "sickened" by the publicity following his disclosure and "incensed" by an offer of $120,000 from a British newspaper for the full story. Although still claiming that the three children are normal, Bevis has also admitted that the technique "is potentially dangerous for the child, and a lot can go wrong."[26]

A rather upset Patrick Steptoe said soon after the announcement, "I am astounded that Professor Bevis would have made this statement. As far as I know, no one in this country or anywhere else has yet succeeded in this technique." (See footnote 26.) Do the children actually exist or not? It would seem they do, but that after an unintentional "slip" in that regard, Bevis and others now wish they didn't. The uproar that Bevis initiated would seem to indicate that research on artificial inovulation may proceed quietly and cautiously for some time yet to come. If success can create such a furor, obviously no one will want to publicize his failures! The birth of a malformed baby could also probably end such research for some time to come.

Dr. Bevis has claimed that the three births did not constitute a "breakthrough" in artificial inovulation, but rather, in a sense, could simply largely be attributed to luck. For, as he said, "So many have been attempted that by the law of averages some have come through." Whether through luck or not, however, it does appear that "the ice has been broken" and that "the door is open." It seems reasonable to predict that sooner or later (probably sooner) another test tube embryo will make it to birth, and for one reason or another, the news will either again "leak out" or will be purposely announced. As Dr. Brackett has said, "It could be done now; it's only a question of when." Once artificial inovulation using original mothers is accepted, or at least firmly established, how long will it take for pressure to mount to attempt the next step—implantation into substitute mothers? And then how long will it take for pressure to mount to take still another step—the establishment of "egg banks" to compliment "sperm banks"? Could we really resist the temptation of making both artificial insemination *and* artificial inovulation more selective? How attractive *would* it be to have extensive control over the choice of sex, physical characteristics, personality traits, and perhaps even intellectual characteristics of *your* children? Is it possible we may even some day take the last step—storing the eggs (as well as the sperm) of famous and/or admired women?

ARTIFICIAL PLACENTAS

Several groups of scientists in Britain, the United States, Russia, and elsewhere are working on the development of an artificial placenta (artificial womb). The immediate motive of the research is to provide a way of aiding prematurely born infants, who are prone to die of breathing difficulties caused by hyaline membrane disease. This disorder kills 25,000 premature newborn infants each year in the United States. Most of the machines under investigation are based on the artificial lung principle. The fetus is placed in a temperature-controlled synthetic amniotic fluid bath at 98.6°F. The umbilical cord passes through a lightly coiled cellulose tube, which bathes it in fluid through which oxygen is bubbled.

Duplication of the composition and concentrations involved in the amniotic fluid is technically much harder than it sounds, since both of these things are variables over the nine-month gestation period. Attachment of the fetus to plastic tubes, the exchange of gases between fetus and environment, the exchange of liquids and solids, and the exchange of hormones and other substances are all quite complicated problems which have not been completely investigated yet. However, many groups of researchers are working on these problems at various universities in the United States and abroad. Many researchers are finding the computer a valuable tool in simulating placenta conditions.[27]

Thus far, fetus survival in an artificial womb in the United States and England has apparently been accomplished for a few days. However, progress in Russia has been far more dramatic. The Russians have used the work of Dr. Daniele Petrucci, a physician from Bologna, Italy, as a springboard for their own research.[28] In the mid-1960's, Petrucci was experimenting with artificially inseminated eggs in artificial placentas—not to produce a nine-month artificial womb baby, but rather to either grow tissues for human transplantations, or reimplant the fertilized egg back into the woman from whence it had come. He succeeded in developing a 59-day fetus, but then terminated its life under extreme pressure from the Vatican. But he was then accused of "murder" by some for the termination! Petrucci, a Catholic, was overwhelmed by the outcry and gave up his work. He was then invited to Russia in 1966, where in a series of lectures, conferences, and meetings he gave the Russians all of his knowledge. The Russians are hard at work on an artificial placenta and have about 250 fetuses growing in artificial wombs at any one time at the Institute of Experimental Biology in Moscow. They have developed their techniques to the point where one of their test tube "humans" grew for six months before it died. They have succeeded in taking a rabbit all the way to gestation and "birth." But apparently all of their human test tube fetuses have been abnormal and have either been purposely terminated at some point or simply reached a state of arrestment of growth. Petrucci is of the opinion that although one of these "monster-like" fetuses could technically possibly become a full-grown, nine-month infant, the Russians will not permit such an individual to emerge from their lab, for both scientific and adverse publicity reasons. While there are apparently technical problems still to be worked out, many, including Petrucci, are predicting the announcement of the world's first *Bionaut* (a human being who spent his nine-month gestation period in an artificial womb) at any time.

CLONING (PARTHENOGENESIS)

The ultimate in human reproduction techniques may very well prove to be cloning. In order to understand how cloning might be accomplished in humans, it might be helpful to review a few simple biological concepts and some historical background information on cloning.

There are two general types of human cells: *body* or *somatic* cells, each of which has a nucleus containing 46 chromosomes; and *reproductive* or *sex* cells (sperm in the male, ova or eggs in the female), each of which has 23 chromosomes. Chromosomes are made up of genes, which are comprised of a chemical known as DNA. A human egg cell could roughly be compared to a chicken egg. The nucleus corresponds to the yolk; the liquid material around the nucleus (the cytoplasm) to the "white"; and the cell wall to the "shell." The cytoplasm contributes nothing to the genetic makeup of the individual. This information is found in the genes. The job of the cytoplasm appears to be chiefly to protect and nourish the nucleus. However, as soon as the sperm swims through the cytoplasm and enters the nucleus, the cytoplasm apparently is "shocked" into sending a chemical message to the nucleus telling it to begin dividing. The "shocked" or "switched on" cell, now containing 46 chromosomes and all the genetic information needed to construct a whole human being, begins dividing (ultimately as many as 100 trillion times) to make that individual. Yet unlike the parent fertilized cell, as division continues, the resultant body cells develop limited and quite specialized creative abilities. Some make only teeth, others only a liver, others only hair, etc. *Each* body cell has the full number of chromosomes and genes necessary to construct an entire new individual, but most of the mechanisms inside seem to "switch off." The genetic components in the 46 chromosomes of a skin cell, for example, are all turned off, except for those that go into making skin. About 80 percent (or even more) of the information in the skin cell is placed in storage somehow. The whole process is called differentiation.

The word *cloning* is a cognate coming from a Greek root word for "cutting." Cloning could simply be defined as being reproduction-without-fertilization, reproduction-without-sex, or even virgin birth. Scientists use the more sophisticated terms *asexual reproduction* and *parthenogenesis* for such a situation.

For years scientists have been fascinated with the idea of taking a body cell and "shocking it" or "switching it on" somehow so that it starts dividing, thus creating an exact replica or duplicate (clone) of the individual from which it came. There seems to be no major theoretical reason why it can't be done. There is also support for the idea from the fact that cloning occurs spontaneously in nature.

It has been known for a long time that an intact, complete worm will result from each of the segments when an earthworm is cut in two. Quite a few other examples of cloning in nature could also be cited, but it's rather interesting to point out that "birds do it, bees do it," and even certain types of fleas do it! For example, some types of plant lice are known to practice parthenogenesis in warm weather and sexual reproduction when it turns colder. Drone honeybees develop exclusively from unfertilized eggs. And, in 1972, a slight tendency toward parthenogenesis in Dark Cornish hens was discovered at the U.S. Agricultural Center at Beltsville, Maryland.[29]

One might intuitively predict that the more complex the life form, the rarer spontaneous parthenogenesis becomes. However, in 1974, at the Jackson Laboratory in Ben Harbor, Maine, bits of bones, cartilage, brain, skin, and muscle tissue were discovered in ovarian tumors in a colony of virgin mice—evidence of spontaneous parthenogenesis. (See footnote 29.) Interestingly enough, the same type of physical evidence has been found in ovarian cysts in human females. And, of course, it might also be pointed out that since identical twins result from the segmentation of a single genotype, they are clones.

The history of attempts to induce cloning date back to the turn of the century. The French biologist, Jacques Loeb, was partially successful in cloning sea-urchins in 1899. Another French scientist, Eugene Bataillon, later repeated the feat with frogs. Both men used a pin prick technique to start an egg dividing, but the division did not proceed to an adult stage. Development was always aborted at some point. Today, high school biology students rather routinely produce tadpoles from unfertilized frog eggs using an improved pin prick method.

In 1952, at the Institute for Cancer Research in Philadelphia, Drs. Robert Briggs and Thomas King replaced the nuclei of freshly-fertilized egg cells of the leopard frog *Rana pipiens* with nuclei from blastula cells.[30] In this way they produced a clone of fatherless, free-swimming tadpoles having the same genetic endowments as the tissue cell donor. In 1956, embryonic clones were produced by using embryonic tissue older than the blastula stage. A few of the 1952 and 1956 clones were allowed to reach maturity, three years being required.

In 1961, building on the pioneering work of Briggs and King, Dr. J. B. Gurdon of Oxford University took an unfertilized egg cell from a South

African clawed frog, *Xenopus laevis,* and destroyed its nucleus with ultra-violet radiation.[31] He then took a body cell from the intestinal wall of another frog and removed its nucleus with the help of a powerful microscope and tiny surgical tools. He implanted the body cell nucleus into the egg cell. Although "tricked" or "shocked" into thinking the cell had been fertilized, the cytoplasm "switched the cell on" and the cell grew all the way to the adult stage—identical to its parent (either male or female) in every way.

Later in the 1960's, cloning was carried out with carrots at Cornell University, tobacco plants and asparagus at the University of Wisconsin, and aspen trees at the Institute of Paper Chemistry in Appleton, Wisconsin.[32] Using the techniques developed in cloning amphibians and plants, there are undoubtedly scientists in the 1970's attempting to clone small mammals. Although no positive results have been reported yet, it seems safe to predict the first successful induced parthenogenesis in mammals sometime before the turn of the century.

Cloning humans will be much more difficult than cloning carrots, frogs, etc., for at least four reasons. First of all, it would be a tremendous challenge to induce the cloning by the cell nuclei transfer technique, for it would be very difficult to get the tiny body-cell nucleus into the egg-cell nucleus. A human cell is smaller than a frog cell. The human egg cell is only about 1/250 of an inch in diameter! Once inside the egg cell, however, theoretically there is no reason whatever why the clone cell shouldn't grow just like an ordinary fertilized egg cell. However, a second problem could develop at this point. There may turn out to be an interaction between the transplanted nucleus and the cytoplasm into which it is inserted. Nuclear-cytoplasmic incompatibility could conceivably lead to abnormalities in development, or no development at all. The subject of cell division is simply not understood yet. Sometimes molecular biologists must feel that the whole question of cell division belongs in the occult arts rather than in biology—for nowhere in their bag of tricks is there any explanation for how cell division is timed or triggered as of yet.

The third problem involves differentiation and could arise no matter how the parthenogenesis is initiated. Our genetic information exists in different developmental "fixes" in the various tissues and organs in our bodies. We would not grow into whole, differentiated bodies unless the genes for eye color were somehow "switched off" in the nuclei of the cells of our fingernails, and the characteristics of the latter "switched off" in the nuclei present in cells in the eye. The "switching off" *may* be the work of special proteins in the cell called histones. These must somehow be counteracted so that we can begin with a nucleus and genotype taken from specialized tissue cells and enable it, when transplanted, to develop into a whole, differentiated member of the species. Alternatively, we might locate some special human tissue in which the degree of differentiation is low and which, therefore, contain a greater degree of totipotency of the original egg nucleus. Nuclei from the early embryonic (blastula) stage of *Rana pipiens* have proven to be totipotent, and this is rather encouraging with regard to this problem. However, it must also be said that the mystery of cell specialization and differentiation is perhaps the central problem in modern biology today, and until there is a breakthrough in this area, there probably will not be a great deal of progress in the cloning of animals and humans.

Actually, that "breakthrough" may not be too far off! There are indications in some recent, parallel research efforts at the University of Indiana and Cambridge, England, that the cell division and differentiation mysteries may both soon be solved.[33] Interestingly enough, both sets of results were reported in the same issue of the same research journal, the March 11, 1976 issue of *Nature.*

At the University of Indiana, Robert Briggs, Ann Brothers, and R. R. Humphrey believe they have now found a cellular agent in egg cells of the axolotl (a type of Mexican salamander or frog) which apparently "throws the master switch" for further development and differentiation. The agent, which may be two or more as yet unidentified proteins, has been dubbed "O+ substance" because it apparently enables the ovum, or egg, to continue development beyond the blastula stage, where differentiation is initiated.

The research work at Cambridge, done by John Gurdon, E. M. De Robertis, and G. Partington, produced quite similar results and conclusions. But, in addition, it was also found that human cancerous cells were also "turned on" when injected into immature frog eggs, indicating that "O+ substance" might also stimulate or at least interact with human cells! Further research at Indiana, Cambridge, and other locations will now be directed toward isolating and characterizing the "O" proteins, and by means of fluorescent antibodies trace their movements within the cell.

The last problem will have to be solved even if a successful cloning technique is developed. It involves the fact that human eggs must be carried in the womb at the present time. Of course, the development of artificial placenta techniques could allow circumvention of this problem, however, and the development of artificial inovulation techniques would offer a way around the nutrient bath or artificial placenta problem.

The investigation of cloning techniques is apparently confined to plants, animals (such as amphibians), and mammals at the moment. No initial work on humans has yet been reported. As we have just seen, the technical problems involved in cloning humans are unquestionably very formidable. And thus there are scientists who pessimistically believe that they will never be solved and that humans will never be cloned. At the other end of the spectrum, however, the most optimistic scientists believe that, with a crash program, the cloning of humans might be accomplished in as little as a few years. In between these two extremes in thought there are scientists who believe it could be done sometime after 2000. As Dr. Robert Sinsheimer of the California Institute of Technology told the 1969 meeting of the American Association for the Advancement of Science in Dallas:

> All that seems needed is the technology to transfer what we already know to be feasible in bacteria, carrot cells, or frogs to man. I feel strongly akin to the physicists who pointed out in the 1930's that the principles required for the release of the energy locked in the atomic nucleus were understood. Here, too, the principles seem in hand.

If and when the cloning of humans is accomplished, there will undoubtedly be once again (as in artificial insemination and inovulation) a tendency or trend develop toward selectivity. The costs involved alone will probably dictate the more extensive cloning of famous people and/or people with "desirable" physical and mental characteristics.

"BABY FACTORIES"

Two hundred years ago the French encyclopaedist Denis Diderot, in a literary version of the future he called "The Dream of d'Alembert," predicted one day the artificial cultivation of human embryos, with their hereditary endowment predetermined. He saw "a warm room with the floor covered with little pots, and on each of these pots a label: soldiers, magistrates, philosophers, poets, potted courtesans, potted kings. . . ."

In 1932, Aldous Huxley wrote a prophetic book titled *Brave New World* in which he envisioned a scientific elite of "Predestinators" who created test tube babies on an assembly line, predetermining in the process every mental and physical characteristic of each baby. Huxley placed his envisioned society 600 years in the future.

The predictions or prophesies of Diderot and Huxley, undoubtedly, seemed like science fiction or "impossibilities" at the times they were written. However, some more recent, similar predictions about the possibility of "baby factories" are far more sobering.

Dr. E. S. E. Hafez of Washington State University speculated several years ago that only 10 or 15 years hence it could be possible for a housewife to walk into a new kind of commissary, look down a row of packets not unlike flower-seed packages, and pick her baby by label.[34] Each packet would contain a frozen one-day-old embryo, and the label would tell the shopper what color of hair and eyes to expect (and other physical characteristics as well) and the I.Q. and perhaps even personality traits of the child. She might also, of course, like to select an embryo containing an egg and/or sperm from a famous person(s). Freedom from genetic defects would, of course, probably be guaranteed! After making her selection, the woman could take the packet to her doctor and either have it implanted in herself for normal birth, pay someone to have the child for her, or leave it at the "baby factory" and have it grown for nine months in an artificial womb. Dr. Bentley Glass, a former president of the American Association for the Advancement of Science, has made similar predictions. In his farewell address at the 137th meeting of the association in 1970, he stated his belief that by the end of the century a fully formed test tube baby will be "decanted" from an artificial womb, ushering in the arrival of the "Brave New World"—more than 500 years "premature."

Nobel geneticist Josua Lederberg has also predicted the ultimate in life insurance—body cell banks. In the event of a person's untimely death, some of his cells, set aside previously and preserved indefinitely in cell-culture solutions, would be taken out of "storage" and used to duplicate that person through the use of cloning, artificial inovulation, and/or artificial womb techniques.

IN CONCLUSION

Admittedly, both the likelihood and/or the imagination-reality times involved in all of the previous predictions are open to debate. And, in all fairness, it should again be emphasized that the technical problems involved are indeed formidable and that many scientists believe they may even be insurmountable. However, at this point in history, very few scientists would deny that anything we have now discussed in the area of human reproduction isn't at least in the area of possibility. Those who believe the predictions will never be allowed to come about because of the pressure of adverse public opinion may be quite surprised to learn that a Harris Poll on the subject published in *Life* magazine, June 13, 1969, revealed a surprising degree of approval of most of the predicted developments. About one out of three people polled approved of AID, artificial inovulation, and "in-vitro" babies with the qualification that approval was based on interpreting these proposals as treatments for childlessness, etc., rather than simply advancement of knowledge.

CHAPTER 2

PHYSICAL MODIFICATION

> It is a good thing Heaven has not given us the power to alter our bodies as much as we would like to and as much as our theories might happen to require. One man would cover himself with eyes, another with sexual organs, a third with ears, etc.
>
> —George C. Lichtenberg

A mirror can be a devastating encounter with instant reality! Which of us hasn't stood before a mirror at one time or another, wishing we might somehow see a different physical and/or mental image. At one time or another, we've all felt that "the grass is greener" on the other side of the genetic fence. It's an age-old daydream that usually just dies of old age. But for the first time in history, we're beginning to take it seriously. When a scientist looks in the mirror today, he sees more than an image comprised of flesh and bones. He sees an image comprised of a specific combination of atoms, molecules, and cells. And as outlined below, many scientists now believe that combination can be drastically changed.

TRANSPLANTED, ARTIFICIAL, AND REGENERATED BODY PARTS

The transplantation of body parts dates back to 1945, when the first attempts to transplant kidneys proved unsuccessful. In 1954, the first successful kidney transplant was performed, between identical twins. The first successful transplant between non-twins was accomplished in 1958. Survival rates of up to five years for kidney transplants are not uncommon today. The first recipient in 1954 lived 8 years.

Although more publicized than kidney transplants, heart transplants have been less numerous, and generally less successful. On December 3, 1967, Dr. Christiaan Barnard, in South Africa, made the world's first human heart transplant into Louis Washensky. He lived 18 days. Dr. Philip Blaiberg, also a patient of Dr. Barnard's, and perhaps the most famous recipient of all, lived almost 20 months. Betty Anick has been the world's longest living heart recipient to date. She survived for over eight years before dying in February of 1977.

Heart transplants have diminished in the United States in the last few years. In fact, in 1975, only the Stanford University Medical School was still doing them on a regular basis. Many doctors apparently feel that if they can't guarantee even one additional year of life to the patient, and only one-third or so of the recipients will live up to five years, the operation is not worth the problems and expense (more than $40,000) involved. The operation has also generated more controversy than the medical profession perhaps anticipated, and even some legal problems.

Transplants of organs other than kidneys and hearts have in general been even more limited and less successful. There have been a few attempts at transplanting either lungs or a heart-lung "package," a limited number of liver transplants, and perhaps 200 or so bone marrow transplants. It is difficult to say how many ovary transplants have been performed, for they are not routinely publicized. However, in late 1972, Dr. Paul Blanco of Argentina publicly announced that a woman on whom he had performed an ovary transplant was three months pregnant. In a public interview, he said that no symptoms of rejection had been encountered and that the pregnancy was proceeding normally. He claimed the baby would be the first after an ovary transplant. There have apparently been no further announcements on the birth, however, which could indicate that the transplant either ultimately failed, or that something else went wrong.

Overall, as of late 1974, 18,325 kidneys, 245 hearts, 217 livers, 35 pancreases, and 34 lungs had been transplanted into patients around the world, according to the American College of Surgeons/National Institutes of Health Organ Transplant Registry. The registry gave no figures for bone marrow, lung-heart, or ovary transplants. The statistics, unfortunately, are misleading. Although one kidney transplant recipient has survived for 18 years, one heart transplant patient more than 8 years, one liver transplant patient 5.1 years, one pancreas transplant patient 2 years, and one lung transplant patient 10 months, as stated earlier, the bulk of transplant patients survive for lesser periods.[1]

Cornea transplants have been performed for years and have been highly successful. However, there is a severe obstacle to success involved in transplanting major organs, over and above the technical, moral, and ethical problems involved. After surgery, the body's immune system goes to work and by producing antibodies attempts to reject or destroy the new organ because it contains foreign tissue. Technically speaking, the molecular cell patterns of the organ and tissue in the body do not match. The molecules on the surface of any cell appear to be arranged in a fairly specific mosaic, sort of as in a giant jigsaw puzzle. There are actually different classes of such mosaics just as there are different blood classes. However, it's far more complex in the case of tissue, and a typing system has not been worked out yet. The mosaics of unrelated people are most different, whereas those of relatives have a number of common features, since many of the genes that control tissue production in such people come from a common source. Identical twins, of course, have identical genes, and thus identical mosaics. Because kidney transplants are obviously more possible between relatives than are heart transplants, the length of survival time had been longer because the rejection problem was less severe.

In an attempt to control and minimize the rejection phenomenon, the doctor will dose the organ recipient with immuno-suppressive drugs. Unfortunately, this drastically lowers the body's ability to reject and destroy *all* foreign tissue, including bacteria and viruses. The patient is now basically defenseless against infections of all kinds and even a minor infection can prove fatal. Well over 50% of the transplant recipients who have died in the aftermath of the operation have succumbed to infection, notably of the lungs. There is a continued danger of infection and rejection and perhaps even an increased danger of cancer since the immune system may be connected with this disease.

Immuno-suppressive therapy is currently an "all-or-nothing" proposition, with the drugs now being used destroying *all* antibodies. The discovery of a selective drug that would suppress only the antibodies that reject foreign heart or kidney tissue, for example, would obviously be a tremendous discovery. Research to develop such drugs is underway, but is unfortunately hampered by an imprecise and incomplete knowledge of how the immune system functions. Research is also underway on using radiation and drugs simultaneously to minimize rejection, and on the possibility of performing bone marrow transplants prior to an organ transplant, since bone marrow manufactures the cells of the immune system. In 1973, Professor John Hobbs of the Westminster Children's Hospital in London publicly announced the first bone marrow transplant from an unrelated donor. The recipient, Simon Bostic, was a two-year-old boy suffering from chronic granulomatous, a rare disease of the immune system. The boy apparently soon died, however, for an article in the May 10, 1976 issue of *Newsweek* claims three-year-old Matthew Ruffer of Bryan, Ohio to be "the first person ever to survive with a marrow transplant from an unrelated donor." Matthew has now survived for over two years after the last of a series of marrow transplants was performed in January of 1975.

Many researchers believe the rejection phenomenon will be successfully overcome within a decade. However, another serious obstacle in the way of extensive organ transplants will still remain—the problem of supply and demand. According to estimates from such sources as the National Institutes of Health, the American Heart Association, etc., the need for just hearts and kidneys for transplants is staggering. About 800,000 people died in the United States in 1975 from heart disease, but there may be about 28 million people currently suffering from one or more major forms of heart and blood vessel disease. At any one time today, there are probably as many as 350,000 people who need new hearts immediately. About 18,000 Americans are currently being kept alive with artificial kidney machines in their homes or hospitals, but as many as 100,000 people annually may need new kidneys!

In terms of supply, however, only a few thousand hearts and kidneys are currently available at any one time, since such organs can only be stored for short periods of time today. Only blood, skin, bones, and corneas can be stored for extended periods at this time. Some people have thus advocated using animals as a more immediate potential source of organs for transplantation. However, to date, the kidneys of chimpanzees, baboons, monkeys, and pigs have been transplanted into men with only very limited success. (The longest living recipient lived for only nine months!) Attempts to transplant hearts into men have been even less successful, with death usually resulting on the operating table. It appears that the rejection problem is even more severe when animal organs are used for transplants, as might be expected. Researchers are predicting, however, that this problem may also be overcome within a decade or so.

If the problems of rejection and availability are solved, transplantation of almost any organ or gland could become commonplace. There may be one notable exception, however—the brain. There is an additional, extremely formidable obstacle standing in the way of brain transplants over and above the problems of rejection and availability. The brain is only one-half of the central nervous system; the spinal cord and peripheral nerve fibers are the other half. The junction between the two halves is extremely complex, and rejoining them would be extremely difficult—so difficult, in fact, that many scientists believe brain transplants are thus "impossible." Research results discussed later in this chapter indicate they could very well be proven wrong in the future. There is one thing on which everyone can agree, though. The brain will undoubtedly be the last body part to be successfully transplanted—if it's ever done at all —and success in this area could not be expected until after 2000.

The development of artificial organs could completely resolve the availability problem involved in transplants, greatly simplify the rejection problem, and dispel many of the moral and ethical objections which have been raised. The extent to which they might be used can be seen in the degree to which simpler body parts have already been employed. According to various estimates, there are about 200,000 Americans with implanted silicone rubber tubes to drain excess cerebro-spinal fluid from the brain, about 100,000 with artificial arteries, 45,000 with artificial heart valves, 70,000 with heart pacemakers, and many thousands more with artificial joints, artificial stirrup bones in their ears, etc. Even large artificial bones are now being implanted.

In 1976 and 1977, many probably considered "The Six Million Dollar Man" and "The Bionic Woman," popular television shows, to be largely science fiction. However, Joan Haas of Florida is living proof that the bionic man or woman may actually not be too far off.[2] Joan probably has more artificial hardware in her body than anyone else in the world at the present time. Some have called her the "six million dollar woman." Between 1965 and 1975, she had 19 operations, largely because of widespread deterioration in her body caused by Bright's disease. She is currently in the process of having at least ten more operations. When the 30 or so operations are completed, she will have had a kidney transplant, spleen removed, thymus gland removed, ovaries removed, and her hip, shoulder, elbow, knee, wrist, and ankle joints replaced by artificial joints.

As of now, artificial organs are still in the research stage, with progress on artificial kidneys and hearts furthest along. There are still many major obstacles in the way of an implantable, internal artificial kidney, but even an external but wearable device would be a tremendous improvement in size and cost over the large, expensive, and immovable kidney dialysis machines now in use. Such machines can currently cost $12,000 or more annually to use. A metal and plastic wearable kidney has recently been developed at the University of Utah which is supposedly small enough to fit into a large handbag. However, further improvements in dialysis efficiency (impurity removal efficiency) should make even smaller units possible.[3] The first wearable artificial kidneys could well be in commercial production by 1980!

Internal artificial hearts could become reality before internal kidneys if several major developmental problems can be overcome.[4] The biggest of these problems involves finding durable construction materials that will last the life of the patient. An artificial heart must survive 400 million flexes in only a ten-year period, but it must also be biologically safe and not cause clotting when connected with blood vessels. Some believe that Teflon, silicone plastics, or some other plastic will eventually satisfy these two criteria. Others believe that collagen, the protein making up about 30 percent of the body, could be the eventual answer. Still others believe the solution may be found in some of the new synthetic rubbers under development.[5]

The size versus efficiency ratio of an artificial heart can also be a problem, since that ratio may have to be related to the age of the patient. For example, a synthetic heart with a satisfactory ratio placed in a baby, or even older child, might have to be replaced later. Interestingly enough, the development of a suitable, long-term power source for a synthetic heart may not be as big a problem as it may seem. Synthetic hearts being tested on cattle are either electrically driven or with an air-driven pump about the size of a grapefruit. Such short-term arrangements are obviously highly undesirable. However, since 1972, researchers have achieved dramatic successes with a series of implantations of nuclear pacemakers. The nuclear pacemakers are expected to last ten years or more without surgical replacement, as opposed to only about two years with the more conventional, external, battery-powered pacemakers currently being used by about 70,000 Americans. The success with nuclear pacemakers has prompted predictions of the availability, within only a few years, of a nuclear synthetic heart, powered by plutonium 238 placed in a container the size of an egg implanted near the abdomen.

Interestingly enough, the first human artificial heart implantation has already been performed—almost a decade ago! On April 4, 1969, Dr. Denton Cooley implanted an eight-ounce plastic heart with an external power source in 47-year-old Haskell Karp. The device kept Karp alive for 68 hours until a natural heart for transplantation could be found. Unfortunately, Karp died 38 hours after the second operation.

There have been no further artificial heart implantations in humans, however, undoubtedly to a large degree because of the controversy generated by the first one. Cooley was subsequently accused of using the artificial heart without the permission of its developers and the hospital involved, and before it had been thoroughly tested on animals.

Glimpses of the apparent imminence of an artificial heart can be seen in the animal testing which has been conducted in the 1970's. In 1972, the National Institutes of Health announced that a nuclear-powered heart of reinforced plastic had been implanted for short-terms in 75 calves, with efficiencies "at least as high as the natural heart." In 1976, researchers led by William Kolff at the University of Utah reported that they had successfully implanted electrically and air-driven synthetic hearts into a series of calves, achieving survival times as long as 122 days. The Utah researchers are very optimistic about their prospects for greatly improved success and believe they can develop a heart powered with a nuclear capsule within several years.

In May of 1976, another research group reported a survival time of 145 days for a calf implanted with a synthetic heart.[6] The seven-month-old calf was a joint project of the Goodyear Rubber Company and the Cleveland Clinic's department of artificial organ research. The calf died of pulmonary edema, after apparently simply outgrowing the capacity of its rubber, air-driven heart, before a proposed booster could be installed. But the heart and its surrounding tissue were still in perfect condition at the time of death.

Overall, it seems safe to predict that eventually over half the body will be replaceable with synthetic body parts. Many of the simpler of those parts are already available. More complex parts such as organs appear to be on the way. Artificial hearts and kidneys will probably be developed first, perhaps as soon as by 1985. Other organs will undoubtedly follow. For example, a research team at Northwestern Memorial Hospital in Chicago is currently developing an artificial lung.[7] If such a device can be perfected, many of the more than 30,000 people who die annually from respiratory ailments such as emphysema, and the more than 75,000 who die annually from lung cancer, might be saved.

There will also undoubtedly be some surprises in the area of synthetic body parts. The first one may be the development of synthetic blood. In the 1960's and early 1970's, Dr. Robert P. Geyer of Howard and Dr. Leland C. Clark and colleagues of the University of Cincinnati reported the successful replacement of up to 90 percent of the blood in rats and dogs with an emulsion of fluorinated hydrocarbons and high-molecular weight polyols (alcohols).[8] Their work, which attracted only moderate interest, has not as yet produced a commercial synthetic blood. But interest in such a possibility was rekindled in March of 1975, when Dr. Jack E. Baldwin of the Massachusetts Institute of Technology announced the discovery of a more life-like blood substitute.[9] A key part of the substitute is a chemical closely related to hemoglobin, known as a capped porphyrin. Baldwin's synthetic blood can carry more than 17 times as much oxygen as an equivalent amount of hemoglobin, but, unfortunately, will only currently work if a chemical toxic to humans (n-methyl imidazole) is also present. Baldwin is working on a modification of the solution which he believes will be safe for humans. Animal tests may already be underway.

The ultimate in body part replacement may be regeneration of those parts. Lizards, lobsters, salamanders, worms, and other forms of life can regenerate body parts. Researchers are hoping ways may be developed for humans to do it also. Maintaining and growing cells and tissue in culture media is at least 50 years old. Since World War II, tissue samples of bone, skin, and testes have been cultured to many times their original size in England and France. Techniques have also been developed for regenerating peripheral nerve fibers leading to muscles and organs, by providing a protective channel through which they can grow. There are currently two ways of doing this. Dead nerves from a cadaver can be transplanted and used as a channel, or autotransplanted nerves can be utilized. Arteries can now be regenerated in a similar way.[10]

Growing organs by the same technique should be far more difficult, for this type of tissue does not start to grow automatically, and there is no way known currently to get it to do so. However, the problem may not be insolvable. As stated previously, every cell in the body of any organism carries all of the information needed to construct new organs, or even the whole body itself. But then, as the original cell from which an individual living thing comes divides again and again, the resultant cells differentiate to produce different parts of the body. The whole process is probably controlled by chemicals such as enzymes and hormones and thus might be reversible in a sense. Might not a stump of an amputated arm or leg thus be "despecialized" by chemical treatment to grow back a complete arm or leg? Could not heart cells be "despecialized" by chemical treatment to grow another heart in a culture or media? There are indications unfolding that it may indeed all be possible.

Since 1973, various researchers have reported the development of techniques for the auto-transplantation of cat muscle,[11] the isolation and in-vitro culturing of heart cells,[12] the development of prototype rat livers and pancreases by culturing isolated cells,[13] and the partial regeneration of amputated limbs in rats through the application of miniscule amounts of electricity to the severed sites.[14] Experiments such as these would seem to indicate that DNA synthesis and cell duplication can be initiated chemically and/or electrically in at least certain types of tissue. This would be a big step toward eventual limb and organ regeneration.

The most amazing, exciting, and perhaps even most surprising regeneration currently being investigated, however, could well be central nerve repair.[15] In the first half of the twentieth century, scientists knew that severed human peripheral nerves could spontaneously regenerate to some extent, but they had no evidence and very little hope that such regeneration could occur or be stimulated to occur in the central nervous system (the brain and spinal cord).

In 1950, William W. Chambers and William W. Windle of the University of Pennsylvania School of Medicine gave a pyrogen (fever producing agent) to several dogs whose spinal cords had been severed. They were looking for the mechanism in the body which produces fever, but were amazed when evidence of nerve regeneration subsequently developed in the spinal cord of one of the dogs. Upon seeing some national publicity describing Chambers' and Windle's work, the mother of a hopelessly paraplegic boy in Georgia volunteered him for a similar experiment. The boy subsequently recovered, adding worldwide publicity and attention to the work of the two researchers.

However, most researchers subsequently claimed that the experiments were not reproducible. The boy's recovery was thought to be simply fortuitous and nerve regeneration in the central nervous system was once again largely thought to be "impossible." There was one researcher who didn't give up, however. Levon A. Martinian of the Academy of Science in Yerevan, Armenia, worked for 20 years in attempts to not only reproduce the work of Chambers and Windle, but improve on it. In 1973, he reported that trypsin and several other enzymes can stimulate such regeneration, and he claimed that after chemical treatments, 40 percent of the 350 rats whose spinal cords he had dissected subsequently recovered from their paralysis, and actually walked!

Windle visited Martinian's laboratory in 1974, concluding that the Russian's work was authentic. Martinian has also visited the United States since, widely explaining his results. Once again, researchers around the world are excited about the possibility of central nerve repair, and they are currently hard at work, attempting to reproduce and confirm Martinian's work. The results could be available at any time. There are already some indications in recent related work that such regeneration is possible. (See footnote 15.)

If central nervous system regeneration becomes clinically available, the potential for its use is staggering. The first benefactors in this country would probably be the 150,000 or so American paraplegics and quadraplegics who already suffer from spinal cord injury, and the 7,500 or so new cases each year. Such repair could also benefit America's 3 million or so victims of head injury, 2 million victims of stroke, and 500,000 multiple sclerosis patients, for whom there is no treatment available at this time. The technique might also provide a major clue or even be a major step toward making brain transplants feasible.

In summary, regeneration techniques are under development in many laboratories around the world. The potential need for regenerated organs, limbs, and nerves is a tremendous stimulus for such work. For example, in addition to the obvious need created by disease, it might also be pointed out that over 100,000 people per year die of injuries and over 400,000 more are physically impaired for life. The regeneration problem is formidable—almost as formidable as the cloning of humans will be. And thus a schedule for the eventual establishment of "limb regeneration clinics," "organ banks," and "organ factories" is difficult to predict. However, many researchers believe that if nature can practice regeneration so routinely, man should eventually be able to do it too. Many believe it will be done within several decades, and possibly by 2000.

GENETIC ENGINEERING (EUGENICS)

With the development of transplant, artificial insemination, artificial inovulation, and cloning techniques, man will be able to alter himself physically (and perhaps even mentally) far more selectively than he can today. However, genetic engineering (eugenics) may be the ultimate technique available in this respect. In order to understand how the process may operate, it is important, and perhaps even necessary, that the reader acquire a deeper understanding of the nature and operation of the cell than was provided in chapter one. The following additional information is provided to help meet that need.

In a very rough way, a man might be compared to a city. A city consists of bricks, boards, and other simple component parts organized into buildings. The buildings are organized into blocks. An organization of blocks makes up the city. Somewhat similarly, a man consists of atoms organized into groups of atoms called molecules. Molecules are the components of cells, and an organized number of cells make up a man. The total number of bricks or boards in a city could probably be estimated or pictured mentally with a little effort, but not the total number of atoms in a man. For example, if all the atoms in an average-sized human being were converted to peas, clouds of these peas were formed, and it started raining the peas, there would be enough peas available to cover the whole earth to a depth of four feet! And there would be enough peas left over to cover 250,000 more planets the size of the earth to a depth of four feet!

A human being starts out as a single cell—the combination of a male cell, called a sperm, and a female cell, called an egg. This cell, called a fertilized ovum, then begins dividing, with division continuing until approximately 100 trillion essentially duplicate cells form. The result is an adult human being. Each cell in the body is too small to be seen with the naked eye, measuring only about 1/250 of an inch in diameter, or about 1/20 the diameter of an average hair. All of the fertilized ova of all the 4.1 billion people currently on the earth would only fill about three-fourths of a gallon container! And yet the 60 to 100 trillion cells from *one* adult, placed end to end, would extend about 6,300,000 miles—about 200 times around the earth!

In a very general way, a cell might be said to consist of three major parts: a cell membrane or wall, a cellular fluid known as the cytoplasm, and a center known as the nucleus. A cell could thus be roughly compared to a chicken's egg. However, the situation is actually much more intricate because each major cell component contains many complex, interrelated subcomponents.

The overall, mind-boggling complexity of the cell can be seen in the exemplar complexity of just one of the subcomponents of the cytoplasm—mitochondria. Although we understand the functioning of mitochondria, the roles of many of the other subcomponents of the cell remain unexplained. Rather amazing and even awesome objects in themselves, mitochondria are small, microscopic specks or bodies that float in the cytoplasm, as many as 1,000 per cell. Each is so tiny that the highest magnification barely suggests that it has a structure. But that structure has now been largely determined. Each mitochondria is made up of as many as 50,000 or more small "tripartite units" or "machines," shaped like "little basketball trophies."[16] The bases of all of the "trophies" fit together to form a membrane, while the "heads" sit on "stalks" that stick up from the bases. The heads pop up and down against the membrane, about 1200 times a minute. With each piston-like stroke, each head manufactures a single molecule of adenosine triphosphate (ATP).

Adenosine triphosphate is the universal power source for all living things. When an organism needs energy for any bodily function, ATP breaks down into simple substances, releasing power as it does so. All of the cells in a human being thus have mitochondria, with one interesting exception: red blood cells. Since the only function of red blood cells is to transport oxygen and carbon dioxide, and they are constantly swept along by the bloodstream, they have no need for power.

The total available energy in three molecules of ATP equals only about that of a single electron flowing through a flashlight battery. However, multiplying about 50,000 "machines" per mitochondria, by about 1000 mitochondria per cell, by about 1200 flexes per minute, by about 100 trillion cells in the body, and dividing by three, yields an astronomical number of electrons—enough, in fact, to give the average human a power rating of about 100 watts, or about as much as a common household light bulb.[17]

Mitochondria might actually thus be thought of as being super-minute, sausage-shaped, floating, pumping power stations which burn fuel (carbohydrates and sugar), produce power in the form of ATP, and also produce waste "ash" (water and carbon dioxide). Extending the analogy, a cell might also be more appropriately compared to a miniature chemical factory than an egg—a fantastically complex chemical factory with mobile, pulsing power stations, an intricate communications network, perfectly trained personnel, a production scheme wherein a few raw materials are turned into thousands of products, and a perfect pollution and garbage-disposal system. A miniature chemical factory that intricate needs an efficient security and boundary system, and the cell has one—its membrane or wall. Although only about several hundred millionths of an inch thick, it functions as an almost perfect "gatekeeper."

Most of the several thousand products produced by the cell belong to a class of chemicals known as proteins. Proteins are the most abundant and the most important molecules in the body. They are the key components of connective tissue, skin, hair, enzymes, antibodies, hemoglobin, etc., and they thus play key roles in the growth, metabolism, defense, and repair processes that take place in a living thing.

A protein is a long, complex molecule, made up of many inter-linked units or parts, known as polymers. Protein polymers are known more specifically as polypeptides. Any polymer might be roughly compared to a chain made up of many links, a train made up of many boxcars, a movie film made up of many frames, etc. However, in the case of a protein, or polypeptide, an analogy of a sentence written in a 20 letter alphabet might be more appropriate. Each of the 20 "letters" is a simpler molecule known as an amino acid. The same 20 amino acids are found in the proteins of all living things. However, different combinations of the 20 "letters" produce different proteins, and thus the diversity found in life, since the number of possible combinations or "sentences" is almost limitless.

Proteins are manufactured on the surface of small, floating bodies in the cytoplasm known as ribosomes. Ribosomes, constituting up to one-fourth the mass of the cell, are thus among the cell's most crucial machinery. Although it's difficult to believe, ribosomes can churn out proteins consisting of thousands of amino acid units about every 10 seconds.

The directions for making proteins in a cell (the blueprint of life) are found in the nucleus in small, flat, stubby, bodies called chromosomes. The nucleus of each cell in a human being contains 46 chromosomes arranged in pairs, 23 from each parent. Other living things have more or less than 46 chromosomes, usually less. A chromosome actually consists of two long strands of polymeric chemicals coiled together in a double helix, much like the strands of a rope. Each strand is called deoxyribonucleic acid, or DNA for short. DNA is actually a very complex polymer made up of 6 types of subunits. However, two of the subunits are simply structural and form a backbone for the molecule. These subunits are sugar and phosphate molecules. The four most important units are molecules known as guanine, cytosine, thymine, and adenine—often simply abbreviated G, C, T, and A. These molecules sort of "hang" from the backbone in sequence. A certain number of these four "letters" spell out the directions for making one protein. Such a sequence is known as a gene. The blueprint of life is written in a four letter alphabet! The same four letter alphabet is used throughout nature, as far as is known. A chromosome might then be compared to a paragraph written out in one long line of sentences. Each sentence would correspond to a gene, but the sentence would be written in a four instead of a 26 letter alphabet.

The two strands of DNA are held together by attractions between C and G and also T and A known as hydrogen bonds. The two strands are actually thus complimentary with regard to these pairs of molecules and fit together like hand and glove in a sense. When a cell divides, the two strands separate and each forms a complimentary strand by picking up G, C, T, and A from the cell environment, through the formation of hydrogen bonds. The original double helix thus reproduces itself, and one of the two resultant double helixes then transfers to the new cell. This process is called replication.

The manufacture of proteins in a cell involves translating the four letter alphabet of the genes into the 20 letter alphabet of the proteins. This process is called transcription. Transcription involves a triplet code, with a specific combination of three units within a gene (e.g. GCT) representing one of the amino acids. Each triplet can also be called a codon. The triplet code has also been called the genetic code. It was not completely defined by researchers until 1967. Each triplet in the sequence of triplets in the gene represents the nature and location of an amino acid in the protein.

DNA is the master architect of the cell—the "boss," so to speak. It orders the manufacture of proteins, although it takes no part in the resultant manufacture itself. Once the blueprint for life is drawn up, or printed on the DNA double helix, the information must be relayed somehow to the ribosomes where the protein manufacture actually takes place. At this point, the DNA turns the job over to ribonucleic acid (RNA). There are two kinds of RNA, messenger RNA (m-RNA), and transfer RNA (t-RNA). Messenger RNA sort of "snuggles" up to the DNA helix and makes a partial copy of the blueprint found there, for example, the directions to make a specific protein. It then passes the blueprint on, chemically speaking, to a series of transfer RNA molecules. Each transfer RNA molecule gathers an amino acid from the cytoplasm and chemically "carries" it to the ribosome. In a sense, transfer RNA acts as a "decoder" for the amino acids that are gathered from the cytoplasm. The amino acids can't read the genetic code and they must therefore be positioned properly by transfer RNA. How each transfer RNA molecule knows how to pick up the proper amino acid molecule from the cytoplasm is a complex chemical story that is not yet completely understood. After passing the blueprint on to transfer RNA, messenger RNA proceeds to the ribosome, and wraps itself around the ribosome. By means of the codons, it then positions the transfer RNA molecules properly and the amino acids are thus strung together in the proper specific pattern, almost like beads. There's an interesting twist in all of this. Every time that thymine (T) is dictated by RNA, RNA substitutes a fifth type of building block known as uracil (U). However, U and A "fit" together just as T and A do and thus RNA can carry out the functions that DNA dictates.

Any changes that develop in chromosomes or genes result in changes in their component DNA and thus in the blueprint or code of life. Such changes are known as genetic defects or mutations. About 99 percent of all the mutations that occur in nature are detrimental or harmful to the life form involved. About 1 percent result in improvement. It is this beneficial 1 percent that is the driving force behind evolution.

Many scientists believe that man is now ready to attempt to produce beneficial mutations in himself in the laboratory, by deliberately altering his own genes. The process by which man may take his own evolution into his own hands has been called *eugenics* or genetic engineering.

Two types of eugenics are currently envisioned, negative and positive. Negative eugenics might be defined as the removal of "undesirable" qualities in man. Those qualities might be undesirable physical traits and/or genetic defects. Positive eugenics could thus be defined as the addition of "desirable" genetic qualities. Most researchers believe that progress in negative eugenics on physical qualities will come first, with successes in negative eugenics on mental qualities and then in positive eugenics to follow later. This is apparently the sequence in which increasing complexity and difficulty will be encountered in research efforts in this area. The possible societal impact of eugenics will be discussed later in the book. At this point, however, let's turn our attention to how it may all be accomplished.

First of all, of course, eugenics can't be practical in a sophisticated, scientific manner unless a basis or foundation in gene mapping has first been established. Various estimates usually place the number of genes in a single human cell at between less than 100,000 to more than 1 million. Some say, however, that there may be as many as 100,000 genes in a single chromosome. In fact, some researchers believe there may be as many as 100,000,000 nucleotide triplets in a single chromosome! This would be several times the number of words in a set of the Encyclopedia Britannica.

At first glance it would seem to be an almost overwhelming job to even begin to classify such a complex cellular system consisting of so many interrelated genes, with regard to their functions and roles. For example, as stated earlier, researchers weren't even sure what genes were until the 1950's, and the genetic code wasn't cracked until 1967. In fact, human chromosomes couldn't even be clearly seen with microscopes until 1970. However, the development of some novel, new chemical techniques, not entirely dependent on the microscope, allowed gene classification to begin in the early 1970's. There have been some rather spectacular results since. For example, using a rather complex technique involving cell fusion, a Yale University research team succeeded by 1975 in mapping all of the nearly 1,000 genes in a certain one-celled bacterium, 2,000 of the nearly 5,000 genes in each cell of the Drosophila fly, and about 100 human genes.[18] In February of 1977, Fred Sanger and co-workers at the Medical Research Council in Cambridge, England, announced that, by fitting together analyses of more than a hundred DNA fragments, they had determined the exact sequence of the 5,375 nucleotides making up the DNA strand of bacterial virus phi X 174.[19] Although the sequence took two and one half printed pages in a journal to describe, the sequencing took less than two years!

Progress in gene classification should be quite dramatic in the next 5 to 10 years. Some believe that well over 1000 more human genes could be mapped in that period. It should be pointed out though, that it will not require the classification of *every* human gene in order to practice genetic engineering. Some genes are obviously of much more interest and importance than others, particularly those that carry a genetic defect.

Any improvements possible in current microscopes could obviously be of great value in gene mapping. Unfortunately, although the light, transmission electron, and scanning electron microscopes now available to researchers have enabled them to "see" chromosomes, genes, DNA, and even two of the largest atoms in nature (thorium and uranium), the images "seen" have lacked detail, specificity, and contrast. In fact, about all that has been "seen" to date have been indistinct, hazy lines, blobs, and dots against a dark background.[20] However, the acoustic, scanning transmission ion and improved electron miscroscopes now under development could allow man to clearly see the component atoms in molecules.[21] Genes would no longer simply look like tiny "blobs." By the late 1970's or early 1980's, it might be possible to clearly discern the four building blocks of DNA (C, G, T, A) and actually "read" the genetic code along a strand of DNA.

Gene mapping, classification, and characterization not only hold the key to genetic engineering, but may also result in tremendous spin-off advances in other areas as well. One of those areas is amniocentesis, a testing technique currently being used to detect certain genetic defects prenatally in over 2,000 pregnant women per year. Amniocentesis involves placing a long needle into the placenta and withdrawing a sample of the amniotic fluid. Some of the cells in the fluid are then examined for any abnormalities in their chromosome content. Various abnormalities are indicative of various genetic defects. If a chromosomal abnormality is found, and is judged to be serious, the parents may consider termination of the pregnancy. However, if the abnormality is not considered to be serious, and is thought to be treatable, drugs or a changed diet may be used in an attempt to correct it. One of the first successful treatments of an abnormal fetus, correcting a genetic vitamin B_{12} deficiency, was reported in August of 1975.[22]

We have been examining the informational foundation that will be needed to conduct genetic engineering. Perhaps it is time to ask the question of how it might be actually carried out. The answer is there are five techniques that have been envisioned for doing it.

The first technique, known as cell fusion, involves fusing two different cells in order to combine their genetic material. The technique was discovered and first applied to mouse cells in 1960 in France, was later refined in the United States and England by 1965, and was then used to fuse human cells with mouse cells by 1967. (See footnote 18.) In the 1970's cell fusion has been used to map genes, as mentioned earlier, and, in 1973, to actually correct genetically-deficient mouse cells in tissue culture. The cells, corrected for an inability to produce a certain protein, were returned to living mice where they functioned normally.[23]

The second technique, known as transformation, involves putting foreign DNA into the nucleus of a cell by either injecting the DNA or placing the cell in a chemical bath containing the foreign DNA. Transformation has already been used on bacteria, simple life forms, and even some mammals, with varying degrees of success. For example, experiments were reported in June of 1974 in which both synthetic and viral DNA were transformed into Indian barking deer cells.[24]

The third technique, essentially a variation of transformation, has been called sperm therapy. It was first reported in March of 1974 at the Memorial Sloane-Kettering Cancer Center, where several researchers previously succeeded in getting live sperm from a mouse to propel itself into somatic or body cells from a hamster, subsequently transferring its genetic material.[25] In November of 1974, Dr. Stephanie G. Phillips of Columbia University reported using sperm therapy to successfully correct an enzyme deficiency in cultured mouse cells. The corrected cells were then reinjected back into the affected mice.[26] Phillips and others believe sperm therapy may eventually be used to correct hereditary diseases in humans, particularly those where the afflicted person lacks a certain specific enzyme.

The fourth technique, known as transduction, involves the use of viruses as carriers of foreign DNA into a cell. Viruses are essentially DNA with a protein coating. When a virus approaches a cell, it attaches itself, dissolves a hole in the cell wall or membrane, and injects its DNA inside. The DNA moves to the nucleus of the cell, where it acts like a "change order," forcing the cell to change the nature of its protein output. Most viruses are detrimental or harmful to the body, but once we understand their role better, it should be possible to produce more and more selective beneficial transductions by introducing viruses that carry pre-programmed instructions into cells. In fact, significant progress has already been made in this respect.

In October of 1971, three biologists at the National Institutes of Health, C. Merrill, M. Geier, and J. Petricciani, announced they had managed to introduce a gene into a genetically defective human cell by means of a virus transduction.[27] The human cells involved were from the connective tissue of a patient suffering from galactosemia, a hereditary disease in which a person cannot metabolize galactose because his cells do not manufacture the necessary enzyme. Galactose is necessary for the normal functioning of life. The three researchers used a virus called bacteriophage lambda which they added to a culture of *Escherichia coli (E. coli),* a strain of bacteria which is found in the human digestive tract and which can metabolize galactose. The virus attacked the bacteria and incorporated the needed information for making the necessary enzyme into its own DNA. Armed with the new information, the virus was then added to the human cells. After three days, the human cells could metabolize galactose, and the next generation of cells, formed by division, also possessed that ability. This indicated that the incorporated gene was transmitted normally. *The first successful genetic engineering experiment with human cells was performed in 1971!*

The last technique, which might be called "micro-surgery," involves the actual, physical entering of a cell and using some type of delicate physical procedure to add to or delete from the genes present, thus producing a desired, specific mutation. Before 1971, it appeared that micro-surgery would probably remain largely undeveloped until well into the 21st century. The extensive gene mapping and classification needed as a prerequisite, the small sizes of a cell and its chromosomes and genes, and the cell disruption produced as a cell and its nucleus are physically entered seemed to be obstacles that could not be overcome for many decades. The only means that seemed possible for conducting micro-surgery were actual surgical techniques, the use of radiation, or lasers. But making such procedures quantitative appeared to be comparable to attempting to crack a walnut with a sledgehammer or producing a painting or a sculpture with a shotgun!

However, in 1971, several years of research at the Stanford School of Medicine and the California School of Medicine at San Francisco began to bear some rather amazing fruit.[28,29,30] First of all, researchers at these two schools succeeded in developing a new micro-surgery technique involving *chemical* scalpels. They found that chemicals known as restrictive enzymes (endonucleases) can be used to cut DNA apart without damaging it irreparably. In fact, the enzymes have an advantage over any physical method that might be used. When they cut DNA, they leave the ruptured strand with "sticky ends" that will immediately attach themselves to any other piece of cut DNA with which they may come into contact.

Secondly, the two research teams found that the new micro-surgery technique would work extremely well with "plasmids." Plasmids are small rings of genetic material floating in the cytoplasm of certain bacteria. About 0.1 to 10 percent the size of a chromosome, they apparently contain a fraction of the genetic material found in the nucleus and specify the synthesis of enzymes that perform several catalytic functions in the cell.

And thus a dramatically new and yet largely unforeseen technique for conducting genetic engineering was born in 1971. In that technique, plasmids are removed from their host bacteria, split open chemically with restrictive enzymes, and then combined with a gene from some other source. The new plasmids are then reintroduced into another bacterium (usually *E. coli*), usually by the use of transformation or transduction techniques. The implanted plasmids change the hereditary characteristics of the resultant hybrid bacteria and act as if they had been there all the time. The hybrid bacteria are known as the "recombinant host," and this new research has thus become known as "recombinant DNA research."

Stanford University biochemist Paul Berg, one of the pioneers in developing the plasmid technique, has said the plasmid experiments are simple enough that some of them could probably be done as high school exercises. In 1972, Berg was preparing to incorporate the genes of a tumor virus into *E. coli,* when colleagues suggested that the resulting bacterium might spread cancer to humans. Berg shelved the experiment. But later, in 1973, when others prepared to do similar experiments, Berg and others issued an unprecedented call for a voluntary moratorium on certain experiments of this kind. After several international meetings in early 1975, and the drafting of stringent guidelines for such research, the moratorium was lifted. This whole situation will be discussed in greater detail in chapter 4 of Part II.

In the last several years, recombinant DNA research has become the "hottest" new area of the biological revolution to develop in the 1970's. Many believe that it could hold the key to answering the critical questions about cell growth, development, and differentiation that are the major roadblocks in the way of developing not only genetic engineering, but cloning, the prolongment of life, and the creation of life as well.[31] At a National Academy of Sciences forum on recombinant DNA research held in March of 1977, Robert Sinsheimer, a California Institute of Technology biologist, said: "I believe science has not taken so large a step into the unknown since Rutherford began to split atoms. . . . The recombinant DNA technology brings us at one bound into a new domain."

Although recombinant DNA research is currently largely in its infancy, there have been some notable results reported since 1974. In 1974, several researchers at the University of California School of Medicine announced the successful implantation of genes from *Staphylococcus* bacteria and from South African toads into *E. coli.*[30, 32, 33] In early 1976, two European and two American research groups reported the successful insertion of a rabbit gene involved in making hemoglobin into *E. coli.*[34] In June of 1976, Ronald Davis and co-workers at Stanford University reported the successful insertion of genetic material from eukaryote baker's yeast into *E. coli,* providing perhaps the first solid and convincing evidence that genes transferred from a higher organism can actually function and express themselves in a lower organism.[35] Davis believes the stage may now be set for a new way of thinking about evolution. It may be possible, he theorizes, that widely different organisms may evolve by occasionally exchanging small amounts of hereditary information through gene transfer. He believes this type of process could be more efficient than random mutations of DNA. Researchers at the University of California, Santa Barbara, have confirmed Davis' work and have also claimed to have preliminary evidence that genes from the fruit fly, Drosophila, can also be transferred to and expressed in *E. coli.* (See footnote 35.)

There is no question but that the work just mentioned is simply the forerunner of even more amazing discoveries yet to come. However, it should also be pointed out that recombinant DNA research is only one of the fronts on which genetic engineering is advancing. As some of the other discoveries that have been occurring in the last two decades are now outlined, the progress that has been made since just 1970 should be particularly noted.

In 1959, a simple DNA molecule was synthesized for the first time. In 1962, Dr. Robert Edgar identified about half the genes in a simple virus and worked out the nature of each enzyme that each gene present produced. By 1967, the genetic code had been worked out. In 1969, RNA was synthesized for the first time, and a research team from Harvard isolated the first gene—from *E. coli.*

In 1970, Nobel Laureate Dr. Har Gobind Khorana and co-workers at the Massachusetts Institute of Technology announced the synthesis of a complete, double-helix, 77 nucleotide unit yeast gene.[36] It was done by assembling "chunks" of the gene produced by enzymes, rather than nucleotide by nucleotide, but it was a tremendous accomplishment nevertheless. Unfortunately, for various reasons, the functional capability of that first synthetic gene could not be tested.

In 1973, Khorana and his colleagues announced the synthesis of a "tyrosine transfer RNA gene," found naturally in *E. coli,* that contained 126 nucleotide units.[37] The gene had the potential to function detectably within a living cell, but lacked control sections that would tell the genetic "machinery" to "start" and "stop." In late August of 1976, Khorana and his co-workers announced the complete synthesis of the 207 nucleotide tyrosine gene—entirely from off-the-shelf chemicals.[38] The first fully functioning, synthetic gene had been made and was found to function correctly in both *in vitro* experiments and in *E. coli!* Khorana claims science is still far from being able to duplicate or synthesize the genes of humans, but it is obvious that his discovery has moved us significantly closer to that day. For example, his work has vindicated the entire hypothetical edifice of modern genetics by experimentally proving that DNA is indeed the genetic basis of life. But beyond that, scientists now have a powerful new tool for studying and experimenting with genes and determining how the structure of a gene influences its function. Khorana's discovery, and its possible ramifications, have prompted Nobel Laureate David Baltimore to say, "Genetic engineering is right down the road." (See footnote 38.)

In 1972, three separate research groups independently reported the partial synthesis of the 650 nucleotide gene in rabbits that directs the production of hemoglobin.[39] In late 1975, another team announced the total synthesis of the same gene.[40] The method that was used can apparently be employed to synthesize many other genes as well, and extensive work is currently in progress in this respect.

The synthesis of genes is well underway. But, it should also be pointed out that so is the synthesis of enzymes and other important body chemicals. In fact, there has been so much progress in this respect since 1969, when the first artificial enzyme, Ribonuclease, was synthesized, that it would be difficult to even summarize. The progress will undoubtedly accelerate also as synthetic techniques improve.[41]

Scientists are also hard at work probing the chemical structure and action of gene "operators, promoters, and repressors"—chemicals that control gene function and expression.[42] Advances in this area would also obviously be important steps forward toward genetic engineering. And, lastly, it should also be mentioned that recent improvements in cell culturing techniques,[43] chromosome sorting techniques,[44] and cell separation techniques[45] have also been important steps forward.

In summary, it is hard to believe that, in a swift quarter century, biologists have made the quantum leap from the identification of hereditary material to its synthesis and transfer between living organisms. Yet this is precisely what has happened. In fact, much of the "leap" actually occurred in the 1970's. With the development of recombinant DNA, transformation, and transduction techniques, the synthesis of the first functioning gene, progress in gene mapping, etc., a new era was born and genetic engineering or eugenics progressed from being only a dream to being a reality! It is true that genetic engineering is still more of an art than a science, but it is genetic engineering advanced enough to have prompted a demand for an unprecedented international moratorium on such work because of its possible consequences.

Dr. James Watson, the co-discoverer of the DNA double-helix, told a House of Representatives science subcommittee in 1972: "The code of life has been cracked and genetic engineering is on its way. . . . Under the magic wand of biology man is now becoming quite different from what he was." It would indeed seem that "genetic engineering is on its way"—at least on its way to becoming the hottest new area of scientific investigation in the latter 1970's! Of course, that doesn't mean that human genetic engineering as an everyday technique is necessarily "right around the corner." However, the dramatic accomplishments of the last few years do seem to indicate that a "threshold breakthrough" has at least been established toward achieving that goal, and that the estimates of from 20 to 50 years hence that many experts have been predicting for its accomplishment may now have to be revised downward. As Dr. Clifford Brobstein, a biologist at the University of California, San Diego, recently stated, "A new genie has emerged from the bottle of scientific research."[46]

ARTIFICIAL AND SYNTHETIC PLANTS AND ANIMALS

In the 1940's and 1950's, biologists made dramatic progress toward understanding cells, their molecular machinery, and the basis of life. That work continued in the 1960's, but another dimension was added. Researchers continued to probe cells and their subunits, but in addition began to attempt to produce their artificial or synthetic counterparts. The progress made in this respect in the last decade or so has been too extensive to completely summarize in terms of specifics, but three of the more notable, recent accomplishments include the development of artificial plant cell wall material,[47] the synthesis of an artificial cell membrane,[48] and the development of a photoelectric "synthetic leaf."[49]

Some researchers have attempted to go beyond synthesizing cellular subunits to actually producing synthetic cells. In fact, there has already been impressive progress in this area.

In 1971, Dr. James Danielli and co-workers at the State University of New

York at Buffalo reported they had "created" a "synthetic" hybrid amoeba, by combining the basic parts of three different amoebas.[50] The hybrid reproduced itself past 20 generations. In 1974, Dr. George Veomett and colleagues at the University of Colorado transferred the nuclei of mouse cells into the cytoplasms of other mouse cells.[51] The reconstructed cells were capable of cell division and growth in tissue culture. In March of 1977, researchers at the Karolinska Institute in Stockholm reported successfully combining parts from the cells of mice and rats.[52] The hybrid cells were capable of cell division. The single biggest barrier to synthesizing artificial hemoglobin may have been hurdled in 1976 at Columbia University when the basic parts of hemoglobin cells were successfully separated and then recombined in yields of 50 percent or better.[53]

Synthetic cell researchers believe their work could eventually provide answers to many or even most of the questions scientists have about cancer, differentiation, genetic disease, biological clocks, and the aging process. One also has to wonder, of course, whether someday genetic engineering, cloning, and synthetic cell research might not be combined to produce a whole new era of "made-to-order" synthetic plants and animals spawned in the test tube. That era is not yet on the foreseeable horizon, but its dawning can no longer be classified as being simply fantasy. As Dr. Danielli has said, "The age of biological synthesis is in its infancy, but it is clearly discernible." (See footnote 50.)

MAN-ANIMAL, MAN-PLANT, AND PLANT-ANIMAL CHIMERAS

At first glance, the idea of man-animal combinations ("humanals"?), man-plant combinations ("humanants"?) and plant-animal combinations ("plantimals"?) seems almost too farfetched to even consider. It perhaps shouldn't be, however, in view of some early progress that has already been made in these three areas.

The successful fusion of human and mouse cells in 1967 has already been mentioned. (See footnote 18.) In 1972, a research team at the Brookhaven National Laboratory announced the production of a whole new species of tobacco by fusing the genetic cells of two existing species.[54] The technique was called parasexual hybridization because it bypasses sexual reproduction. The researchers involved believe the method could eventually be applied to many forms of plants and perhaps even further into the future to mammals. In June and July of 1976, three separate research groups reported the successful fusion of several combinations of plant, animal, and human cells.[55] A team at the Brookhaven National Laboratory claimed to have successfully fused human cells with tobacco cells and to have kept the resultant hybrid cells alive for about six days (without reproduction). A second group of researchers, headed by Dr. James Hartmann at Florida Atlantic University, reported successfully fusing rooster red blood cells with tobacco cells to produce hybrids that survived up to five hours. The third research group, at the Hungarian Academy of Science's Biological Research Center, has successfully fused human cancer cell nuclei with carrot cell nuclei.

At this point in time, fused cell research is also in its infancy. It's difficult to say if and when the problem of one of the sets of genetic material being lost upon successive divisions of the hybrid cell can ever be solved. But if it can, the results could be staggering and limited only by the imagination. For example, James Hartmann has predicted the possible production of a meat-quality cell capable of producing its own food through photosynthesis, and perhaps eventually even animals able to sustain themselves on fertilizer and sunshine. (See footnote 55.)

It seems quite likely that, as time goes on, there will be an increasing temptation to fuse the cells of humans and animals, or mix their DNA in other ways, in attempts to produce man-animal mixtures (chimeras) in varying proportions. The precedent has already been set in the man-mouse cell fusion work mentioned earlier. The specific purposes of such research could be to provide a source of organs for transplantation, create a unique labor force, etc. As Nobel Laureate Joshua Lederberg said several years ago,

> Human nuclei, or individual chromosomes and genes, will be recombined with those of other animal species; these experiments are now well under way in cell culture. Before long we are bound to hear of the tests of the effect of dosage of the human twenty-first chromosome on the development of the brain of the mouse or gorilla.[56]

IN CONCLUSION

The physical modification of man is indeed "on its way." However, its final destination is difficult if not impossible to predict at the moment!

CHAPTER 3

MENTAL MODIFICATION

The mind is its own place, and in itself
Can make a heaven of hell, a hell of heaven.
—John Milton, *Paradise Lost*

In appearance the brain is a three-pound mass of folded putty and sloppy jelly, a rather unlikely-looking candidate for being the master organ of the body, in control of bodily functions, thought, memory, and emotions. Thus it was that many through the ages believed that the brain was only a "transmitter," broadcasting commands from some invisible controlling soul or spirit. Many have even believed that the seat of the mind was actually the heart or the stomach. And thus sayings arose such as, "I haven't got the heart or the stomach for it," "He has a broken heart," etc.

In the sixteenth, seventeenth, and eighteenth centuries, surgical dissections revealed the existence of four distinct components in the brain: cerebral cortex, thalamus, limbic system, cerebellum and brainstem. It was discovered that the cerebral cortex consists of two interconnected hemispheres, comprising 80 percent of the brain. The other three components are buried within the cerebral cortex. A rough correlation was found between certain areas of the brain and certain bodily functions.

In the nineteenth and twentieth centuries researchers have found that each component and area of the brain controls certain functions. The limbic system appears to be the seat of the emotions. The cerebellum and brainstem work below the level of consciousness and control voluntary and involuntary muscular coordination. The thalamus and cerebral cortex together integrate incoming sense data. The cerebral cortex is the brain's most elaborate center, governing such complex conscious functions as vision, speech, etc. Here sensations are registered and many voluntary actions initiated involving the thought processes. Specific areas of the cortex control specific functions.

THE ELECTRICAL CONTROL OF THE BRAIN

In the last century, attempts to "cure" various physical and mental disorders through lobotomies and other types of brain surgery often ended in death or severe bodily impairment. Today, however, although brain surgery has become a rather routine, extensively-used medical technique to treat physical problems, it has been used much more "quietly" to treat mental problems. Thus, the extent to which brain surgery has progressed and is currently being used to treat behavioral disorders is difficult to assess, for there have been very few announcements from researchers in this area in recent years.[1]

There still seems to be a stigma attached to "cutting" into a person's head and brain. Interestingly enough, however, there seems to be much less aversion to implanting electrodes into the brain and electrically stimulating it. Such research has been rather extensively reported and publicized.

For almost 50 years, scientists have been electrically stimulating the brains of both animals and men, passing electrical impulses and radio waves into various parts of brains and measuring the results. There have been many such studies conducted, but perhaps the best known are those by neurosurgeon Dr. Jose Delgado. For nearly two decades he has probed the secrets of the electromagnetic waves or "brain waves" given off by the brain. By transmitting radio signals to electrodes implanted deep within the brain and interfering with existing brain wave patterns, he has halted a charging bull in its tracks, caused female monkeys to lose all interest in their young, and even induced cooperative attitudes in previously recalcitrant human patients during psychiatric interviews.[2] Delgado has also successfully programmed a computer to stifle brain waves produced in one area of a chimpanzee's brain by radioing to another part of the brain a control signal that turns them off.

To date, such experiments have usually seemed to be at least somewhat "out of context," for researchers still have very little idea of exactly where in the brain control signals should be delivered to get an exact response. For this reason, the experiments are usually plagued by a lack of predictability, and there has been a pronounced inability to create what might be termed a "robot performance." Nevertheless, the ultimate promise of the research remains quite clear, for there have been definite indications that movement, affection, aggression, pleasure, anxiety, fear, violent behavior, and their opposites can be aroused and to at least some degree controlled through the electrical stimulation of the brain.

Many brain researchers today claim that brain stimulation can only trigger or intensify emotions or motivations such as those just mentioned, and that it cannot "create them from nothing." Delgado, for example, has said that human character cannot be remolded with stimulation techniques because "you can only induce what is already there." He thus maintains that electrical stimulation cannot produce permanent "programmed" personality changes. But to what extent is such a claim actually a matter of interpretation and thus open to debate? Are triggering brain messages and creating them really two different things? And, even if they are only being triggered today, is it possible they could be created in the future?

Electrical stimulation of the brain has already been used in rudimentary attempts to modify behavior.[3] However, such attempts have been (and are) receiving very little publicity, probably because of the serious emotional and ethical issues involved. The path to the eventual acceptance of behavior modification using such techniques could conceivably lead from two by-products of similar research (one expected and one unexpected), which almost everyone will label as being "beneficial advances."

The expected by-product is electrical brain stimulation to prevent pain.

> As James S. awakens, his arthritis is acting up again. He reaches for a battery-charged box on his bedside table, switches it on and places it near his chest. An electrical charge generated by the battery pack stimulates receivers implanted in his upper chest, and then runs along wires implanted under the skin of his neck and up into tiny electrodes implanted in his medial brain stem. Several minutes later his joints stop hurting and he remains pain-free for the rest of the day.
>
> —*Science News,* November 22, 1975

Before 1970, the above scenario would have been considered to be essentially science fiction and "impossibility." However, in late 1969 David Reynolds of the Stanford Research Institute discovered that electrical stimulation of the medial brain stem can inhibit pain. (See footnote 3.) The medial brain stem, a continuation of the spinal cord, is located deep in the middle of the brain. Pain relief has subsequently moved from that discovery to the near commercial production of devices such as those just described—in less than a decade! The devices are already being tested on human patients, apparently with good results.[2,4] It would appear to be only a matter of time before electrical stimulation of the brain to relieve pain becomes widely used. One can't help but wonder how big a first step such techniques might be to eventual pleasure-wiring, happiness-wiring, etc.!

Interestingly enough, pain relief research has produced an unexpected bonus of great potential significance. In 1970, a research team at UCLA discovered that the pain relief achieved through electrical stimulation could be reversed with naloxone, a morphine antagonist.[5] The discovery suggested that the brain contains some natural chemical pain inhibitor, similar to morphine, which has the same affinity for nerve cells and can be blocked by the same agents (e.g., naloxone). In early 1976, several researchers reported confirmation of the presence of such an inhibitor in the brain, and also reported identifying and synthesizing it. The substance was named "enkephalin." In May of 1976, another research team, at the National Institutes of Health, reported finding a similar inhibitor in the blood stream.[6] They named it "anodynin," from the word *anodyne,* meaning a medicine to relieve pain. It is beginning to appear that there is a whole series or network of pain inhibitors in the body which may eventually be manipulated to relieve pain.

The unexpected by-product of electrical brain stimulation research, which, like pain relief, could also eventually help bring respectability to behavioral modification, is an artificial eye. Since 1974, researchers at the University of Utah and the University of Western Ontario have succeeded in electrically stimulating the visual cortex of blind patients to the extent that they can recognize simple patterns and letters and even mentally "read" mental images faster than they can tactile braille.[7,8] The stimulation is accomplished by sending light images from a television camera through a computer, which changes them to electrical signals and then sends them into either electrodes implanted in or on the visual cortex. Wires from the electrodes or electrode wafer emerge from above and behind the ear. Eventually the researchers hope to perfect a functional artificial eye in the form of a tiny television camera mounted in the eye socket of a blind person. The camera would send light impulses to a miniature computer mounted in an eyeglass frame. The computer would convert the light impulses into electrical impulses and send them to electrodes mounted in the brain. The impact of such a discovery would be tremendous. There are an estimated 350,000 legally blind persons in the United States. However, fewer than 20 percent read braille and only about 10% can get about with a cane or seeing-eye dog. (See footnote 7.)

THE CHEMICAL CONTROL OF BEHAVIOR, MEMORY, AND INTELLIGENCE

There seems to be little question that at some time in the future man will be able to quantitatively produce behavior modification through the electrical stimulation of the brain. However, the chemical route could ultimately prove to be even more attractive, for it holds the additional promise of leading man to memory and intelligence modification as well. In order to fully understand and appreciate the possible long-range potential of the chemical control of the brain, it might be helpful to back up briefly and examine the nature and functioning of the brain in somewhat greater depth than we did at the beginning of the chapter.

The brain sends and receives impulses to and from various parts of the body through a network of nerve fibers called the peripheral nerve system. The main trunk lines consist of smaller nerve fibers which emerge from the brain and spinal cord, bundle together into the large cranial and spinal nerves and then divide and subdivide in the outer reaches of the body so that, finally, single nerve fibers reach into every area of the body. The peripheral nerve system is extremely complex.

The central nervous system (CNS) is comprised of the brain and spinal cord. These two components (and the peripheral nerve system as well) are in turn made up of nerve cells. There are about 10^{14} cells in the body, with about 10^{12} of them in the CNS. The brain contains approximately 10^{10} (10 billion) nerve cells. Actually there are two general classes of nerve cells, neurons and glial cells. Glial cells always outnumber neurons. For example, two-thirds or more of all brain cells are glial cells. They are generally thought to provide a structural matrix for, and supply chemicals and energy to, the neurons, although most researchers concede their role may be far more complex.

Two types of neurons apparently do most of the work in the CNS. Sensory neurons collect and relay impulses received at reception sites such as the skin, nose, ears, eyes, organs, etc., to the brain. Motor neurons carry impulses back to "working cells," usually muscle cells.

Each neuron in the CNS consists of three major components or parts: the cell body, a number of fibers known as dendrites, and a single, longer fiber known as an axon. A neuron may have from one to hundreds of dendrites, but, except in rare exceptions, has only one axon. Axons are often surrounded by a sheath of fatty material known as myelin, and might be compared to a "wire inside an electric cord."[9] An axon is only about 0.001 inches in diameter, but it may be several feet long, as in the axons of spinal nerve cells. It may or may not branch at the end, ending in one or more bulb or knob-like structures known as synapses or synaptic terminals. The synapses contain, among other things, mitochondria and a great number of small bodies known as vesicles. The space or junction between a synapse and a dendrite or cell body is known as the synaptic cleft and is 200 to 300 angstroms (0.0000008 to 0.0000012 inches) wide.

With a few exceptions, neurons might be considered to be elaborate devices for transferring nerve impulses from one to another at synapses—but in one direction only. A neuron receives the impulses at synapses located among its dendrites (or on its cell body), the dendrites pass them to the cell body, and the cell body passes them out through its axon, which in turn terminates at another dendritic cluster and/or at the cell body of another neuron. And thus, in a sense, dendrites receive impulses and axons transmit them. Each neuron may make as many as 10,000 or more connections with neighboring neurons by means of synapses connecting it with those neurons. In fact, some believe the figure can go as high as 100,000 connections!

All neurons in the brain are interconnected, with an astronomical number of *different* circuits or wiring diagrams thus obviously possible. Although there exist certain evolutionary modifications in neuron structure between species, the major difference between various forms of life appears to be in the complexity of neuron organization. "Higher" organisms possess more complex nervous systems that permit vastly greater degrees of integration and information storage, and thus greater plasticity of behavior. In man, a precise, complex, and selective organization of interconnected neurons is apparently present and operative at birth. It is possible that connections between neurons can be uncoupled and/or new ones formed, thus prompting some researchers to speak of "plastic neurons." However, the mechanism and circumstance of these changes, if they occur at all, are unknown. Unfortunately, for obvious reasons "circuits" in the brain cannot be easily traced and it is virtually impossible to monitor changes in them.

Until the 1950's, the CNS was thought of almost exclusively as an electrical system. Older textbooks compared the CNS to a telephone switchboard. More recent textbooks still compare it to a computer. It is true that nerve impulses are electrical in nature. Minute electrical currents do flicker from cell to cell due to the movement of differing concentrations of different ions (charged atoms), which produce differences in electrical potentials. The details of these processes have been rather well worked out for some time now. However, in the 1950's and 1960's, researchers discovered that nerve impulses are more than just electrical messages. Electrical transmission down a nerve is now known to be replaced by chemical transmission at the synapse. The brain thus actually works on chemical as well as electrical principles!

At the synapse, the nerve impulses prompt the release of chemical "transmitters," apparently from vesicles. The transmitters pass through the presynaptic membrane, cross the synaptic cleft, and then pass through the postsynaptic membrane of the second neuron, where they combine with "receptor" chemicals, subsequently regenerating new electrical nerve impulses and thus stimulating the second neuron.

The synapse has been likened to the gate of a computer. Release of transmitters from the presynaptic cell can either inhibit or trigger the firing of the postsynaptic cell, depending upon what the chemicals are that are involved and what synaptic potentials are set up. Some transmitters are excitory, or "yes" messages, others inhibitory, or "no" messages. The postsynaptic cell apparently adds up all the yes and no messages and sends a signal outward down the axon, if the sum exceeds a certain minimum amount.

And thus, in a very real sense, it is at the synapse where "the action is." The axon is concerned only with the transfer of "bits" of information through the whole system, but the synapse, through the use of transmitters, differentiates between patterns of "bits" and ultimately accomodates to experience, thus forming the basis for intelligence, emotion, and memory. Because of the importance of the synapse, some have said the brain could actually be thought of as "a system of synapses connected by neurons."

Much of the brain research taking place today is thus obviously centered around the identification of transmitters. There has been significant progress made in the last two decades.[10] Chemical transmitters known as serotonin and dopamine were discovered in the 1950's. Dr. Ulf von Euler of Sweden, Dr. Julius Axelrod of the U. S., and Sir Bernard Katz of England shared the 1970 Nobel Prize in medicine for their work in identifying two more transmitters, noradrenaline and acetylcholine, in the 1960's.[11] They also identified the role of these two transmitters at the synapse, information which led to the development of anti-depressant drugs and L-Dopa for treating Parkinson's disease. Although about 350,000 people are still admitted to mental hospitals in the United States each year, this level is no greater than that of 1947, thanks to anti-depressant drugs.

All of the transmitters studied and identified to date have been found to belong to a class of organic compounds known as amines. The transmitter amines are known collectively as the biogenic amines. An amine is a chemical compound which has one or more nitrogen atoms in its basic structure. The nitrogen(s) may be part of a chain-like molecule or part of a five or six-membered ring structure in which the other members are carbon atoms. Almost every psychoactive drug which society is now using and/or abusing, legally and illegally, is an amine and often bears a striking similarity to the biogenic amines. It seems quite likely then that heroin, barbiturates, LSD, etc., produce their mind-altering effects by disrupting, suppressing, and/or over-stimulating the synthesis, storage, and/or metabolism of the normal transmitter biogenic amines in the vesicles and synaptic junction. It is thus quite possible brain research could help solve the drug problem, and vice versa.

We have only scratched the surface, however, with regard to the nature and role of chemical transmitters and receptors. For example, little is known about how the electrical nerve impulse activates the release of transmitters or how the transmitters reactivate the impulse beyond the synaptic cleft. The answers to such questions will not come easy, mainly because the extremely small size and complex nature of the synapse makes it a very difficult object to study. However, it should also be pointed out that brain researchers have recently added a powerful new weapon to their arsenal in this respect.

Synaptic vesicle action is currently being studied with an electron microscopy technique first used in the mid-1970's at the University of California Medical Center at San Francisco.[12] The technique utilizes the rapid-freezing methods that stop sub-cellular action and "freeze" the action of vesicles within 0.5 milliseconds. With further development, the technique may soon allow investigators to directly study the changes involved in alterations of the nervous system, by actually "seeing" packets of chemicals being released from nerve endings. Some researchers believe it is quite possible that the brain responds to experience by changing the efficiency of such release. With electron microscopy, synaptic structure and behavior are now being greatly clarified, almost down to a molecular level.

Overall, the results of our current, rudimentary knowledge have already been spectacular. But the future should prove to be far more so, for the synapse is the likely site of the coding or information transducing properties of the brain. If the chemical transmitters and receptors present at various synapses can be fully identified, a way of influencing behavior and mental performance, while leaving other aspects of brain function perhaps completely unaffected, may be found. If the transmitters and receptors governing neurons associated with such functions as moods, pleasure, appetite, sex, sleep, etc., turn out to be specific, and chemical methods can be found for selectively interfering with or enhancing their metabolism, then fairly precise behaviorial modifications might be brought about. In other words, we might then make a science out of what is now only an art! Man already has an arsenal of pain relievers, sleep-inducing agents, depressants, anti-depressants, and hallucinogenic drugs at his disposal. But the metabolism and role of such drugs in the brain is usually not understood today. Even the role of aspirin is still largely a mystery! With a further understanding of brain chemistry, however, far more sophisticated and selective behavioral drugs will undoubtedly be developed. As Dr. Axelrod has said, "This is the wave of the future . . . this is the frontier." (See footnote 11.)

As stated earlier in the chapter, a chemical understanding of the brain could lead to more than simply behavior modification. Since intelligence and ability to learn are only loosely connected with brain size and mass in nature and man, it would seem to be logical to state that it must be the number, size, and degree of development of synaptic terminals in the brain which determine the degree of intelligence and learning ability of a living thing. In humans, educators have long maintained that there is a direct correlation between intelligence and experience and environment. In June of 1972, a 10-year research study conducted at the University of California added strong support to that theory from a new angle—the cellular level.[13] The study revealed that laboratory animals subjected to an increased number and variety of experiences had heavier cerebral cortexes, increased glial cells, and fewer, but larger and more fully developed synaptic terminals than "uneducated animals." It may thus well be that the easiest and most practical way to modify intelligence and learning ability is to stimulate a developing brain with experiences. It might also be the most widely applicable method, since most of the children in the world do not have an optimal childhood. In fact, simply increasing the protein intake of infants and children in many parts of the world could raise intelligence levels, since about 80 percent of all brain growth takes place in the first three years of life.

"Rewiring" the brain with injections of chemicals rather than experiences may prove to be far more difficult. There are no chemical agents yet known, or on the foreseeable horizon, that can effect dramatic, permanent improvements in intelligence. That is not to say it can't be done, however. In fact, there are some indications that, at a minimum, lesser improvements are possible. For example, stimulants such as amphetamines, caffeine, strychnine, etc., taken when we are "too tired to think," could in a sense be said to be temporary intelligence enhancers. It thus seems likely that other, improved chemicals may also be found or developed. At the moment, various researchers are isolating, identifying, synthesizing, and investigating various hormones in this regard, and are testing them on rats.[14] Although the progress achieved to date has been slow and far from spectacular, it does prompt a prediction that a specific chemical enhancer will eventually be found. However, when this will happen, and how much enhancement may ultimately be possible are still unanswerable questions. The key to the whole quest could be found in the development of chemical memory agents.

We are just beginning to unravel the mysteries associated with memory. After 30 years of searching for the part of the brain where memory might be stored, with no success, most researchers now believe memory is a phenomenon that must be associated with the brain as a whole. Some researchers believe that the memory process is electrical in nature. For example, Dr. E. Roy John of the New York Medical College believes that he has found increasing evidence in the last 20 years that memories consist of electrical patterns that can sweep repeatedly through whole populations of nerve cells in the brain.[15] He believes memory thus has an electrophysiological basis and results from the pattern firing of neurons in the brain, rather than the firing of individual neurons. John calls the patterns "engrams," and claims that the original pattern for producing an engram is fixed by an original experience. In January of 1975, he reported new evidence to support his theory.[16]

There is another theory about memory which has also been gathering support and momentum in the last 20 years or so—the idea that engrams may actually basically be a chemical phenomenon. John claims that his evidence complements, rather than contradicts, evidence that memory is chemical. "Memory storage is chemical," he has said, "but the manifestation—the recall —of the stored memory is electrical. (See footnote 16.) Others are not so sure, however, and it currently appears that a lively controversy over the possible degree of divergence in the two theories will continue for some time to come.

The chemical basis theory dates back to the late 1950's when Dr. James V. McConnell of the University of Michigan performed a series of experiments with planarians (flatworms), a small worm found at the bottoms of springs, ponds, and streams in most parts of the world.[17] McConnell trained planarians to respond to a flash of light and then cut them in half. When the ends had regenerated, both remembered their lessons. When ground-up, trained planarians were fed to hungry planarians, the cannibalistic worms acquired some of the trained worms' knowledge. However, adding a small amount of an enzyme that destroys ribonucleic acid (RNA) to the water in which the planarians lived resulted in little or no memory improvement. The experiments seemed to indicate that memories can migrate and that the migratory carrier or agent almost had to be a chemical. RNA was tabbed a likely candidate as being that chemical. (Note: The reader may recall that RNA copies the "blueprint" for making proteins from DNA in the cell nucleus, carries the blueprint to the ribosomes in the cytoplasm, and then directs and helps synthesize those proteins.)

The realization that the memory process may be chemical in nature is only a decade or so old. It has led to a growing optimism that eventually an understanding of the whole process may be possible. At the moment, however, there is only confusion. For example, since the late 1950's, experiments too numerous to mention, specifically to elucidate the role of RNA in the memory process, have led to the formation of two theories in that regard. The data currently available seems to support both equally well. One theory states that learning and memory are actually contained within the RNA molecule. Just as DNA stores and "remembers" an organism's "ancestral memories" in coded chemical form, RNA might encode or "remember" an organism's own personal memories. In other words, RNA might be the chemical "tablet" on which the fingers of experience write, by changing the chemical code carried by the RNA molecule. The other view is that learning and memory are almost exclusively nerve cell transmissions, with RNA perhaps playing a role somehow.

Both the RNA question, and the overall chemical basis theory itself, are currently clouded by several "loose ends" that no one has been able to tie together yet. First of all, there is the rather controversial work of Dr. Georges Ungar, a pharmacologist at the Baylor School of Medicine, to consider. In the early 1970's, Ungar and others claimed to have performed many experiments similar to those of McConnell's earlier work, but with rats and goldfish rather than planarians.[14,18] They also claimed to have extracted three "memory molecules" (including a compound which Ungar named "scotophobin") from the brains of the "trained" rats and goldfish, which could be injected into "untrained" subjects to produce similar behavior. Ungar's work has proven to be difficult to replicate, and has thus drawn extensive criticism from several sources.[14,18,19] However, in the spring of 1975, three researchers at Northwestern University announced positive results in work similar to Ungar's, thus helping to keep his claims alive.[20] Another "loose end" can be found in the fact that a large mammalian neuron synthesizes about a third of its protein every day. But much of that protein leaves the cell via the axon, and its ultimate fate is still an unresolved mystery. Two more "loose ends" involve the claims of some researchers that glial cells are somehow involved in the memory process, and the claims of yet others that the pituitary gland is involved. (See footnote 14.) The latter claim received strong support in January of 1976, when several researchers reported that the injection of certain amino acid sequences present in several pituitary hormones, into human volunteers, improved learning and visual retention, reduced anxiety, and showed some promise for treating memory disturbances.[21] Similar results had previously been obtained in animals.

Although understanding of the memory process is still at a rudimentary stage, it seems safe to predict that more and more of the mystery will be unraveled as time goes on. Certainly we will know in time whether memory is an electrical process, chemical process, or both. If the process turns out to be largely or completely electrical, memory-control may be somewhat difficult. But if the process turns out to be chemical or controlled by chemicals such as DNA and RNA, and those chemicals can be sufficiently characterized to be artificially synthesized, it may be possible to enhance, edit, or erase the memories of individuals! It may be as easy to eventually implant artificial memories in people as it apparently has already been to implant them in planarians, rats, and goldfish. When all of this may be done is not predictable at the moment. It would be naïve to expect quick results because of the difficulties involved in the research and the complexity of the problem. In fact, Dr. Ungar has stated that he "doubts that we will break the chemical code for memory in my lifetime." (See footnote 18.) It should perhaps be pointed out, however, that Dr. Ungar is no longer a young man! And, it might also be pointed out that there already seems to be an unusual excitement and even optimism in the air among those engaged in such research.

DISEMBODIED BRAINS, HEAD TRANSPLANTS, BRAIN TRANSPLANTS

At first glance, the subject of disembodied brains, head transplants, and brain transplants seems too ludicrous to even discuss. However, it should be pointed out that many researchers around the world don't feel that way. They have been hard at work in this area for some time now, and they take their work quite seriously! For example, in the late 1960's a Russian medical researcher announced that he had successfully transplanted the head of a dog, still attached to a portion of its body that included the forelegs, onto the entire body of another dog.[22] The chimera supposedly lived for several days.

The leading United States researcher in this field is probably Dr. Robert J. White, a neurosurgeon associated with Western Reserve University in Cleveland. In 1967, Dr. White reported that he had successfully kept more than 100 monkey brains alive, outside of their skulls, for up to "several days" by hooking them up to either a heart-lung machine or the circulatory system of a sedated, donor monkey. (See footnote 22.)

In 1973, Dr. White reported that he and co-workers had refined their techniques to the point where they had succeeded in transplanting the entire heads of 10 monkeys to the bodies of others and had kept the resultant chimeras alive as long as a week.[23] Although the lower bodies were "dead" for all practical purposes, because there were no nerve connections between the head and body, tests showed that the transposed heads could see, smell, taste, hear, move their faces, close their eyes, and experience pain!

White has said he believes isolating a human brain or even head might actually be easier than isolating that of a monkey, for the human head is larger and the arteries and nerves would thus be easier to operate on. And, he also believes that such an isolated system might be valuable, for example, as sort of a "hormonal computer," in which chemical input and output could be studied without the rest of the body interfering by being in the circuit.[24]

However, he has also said that he does not believe that either brain or head transplants will be feasible in the foreseeable future. He backs up that contention by first of all pointing out the extreme difficulty there would be in trying to keep oxygen deprivation under four minutes in such operations, and thus in preventing brain death. Secondly, he points to the extreme complexity involved in trying to rejoin the brain and spinal cord. If they were not rejoined, White says, the result would be "freakish—a human head attached to an unresponsive body."

The situation with regard to transplants actually may not be as bleak as White paints it, for he has also conceded that it is possible that cooling techniques could be developed in the future which could circumvent the oxygen problem. And it should also be pointed out that since White expressed his reservation about rejoining the central nervous system, there has been a dramatic revival of interest in central nerve regeneration.

If disembodied human brains and heads are already possible, and brain and head transplants may eventually be possible, why hasn't there been any research reported in these areas? With regard to disembodied brains, Dr. White has said,

> We could keep Einstein's brain alive and make it function normally today. . . . It can be accomplished now with existing techniques. . . . I will not because I haven't resolved as yet this dilemma: Is it right or not? . . . There are still too many unresolved problems for me to use a human brain. . . . We are not ready for it. Neither socially nor psychologically nor sentimentally. . . . The passage of a considerable period of time may be required to lessen the impact factors.[22,23]

One can't help but wonder, however, whether White and/or other researchers may not eventually change their minds about man not being "ready for it."

The haunting question also remains of what might constitute a "considerable period of time."

MAN-COMPUTER AND MAN-MACHINE CHIMERAS

Science fiction has long depicted individuals who were either hooked to computers and machines, or were able to control them from a distance with brain waves. In the last decade or so, many of the discoveries and advances which we have now discussed have been injecting small, but increasing amounts of potential reality into such "fantasy." There are also three additional areas of development currently unfolding, which we haven't yet discussed, that are also beginning to provide massive injections!

First of all, some researchers believe they now have solid experimental evidence that there is a "cerebral asymmetry" or "hemispheric dominance" present in the brain, apparently present from birth.[25] They claim there may actually be two minds present in the two hemispheres that operate interrelatedly in some ways and independently in others.[26] Overall, the left half appears to be more logical, analytical, scientific, and technological, whereas the right half may be more creative, emotional, humanistic, and holistic. Asymmetry research is the newest and perhaps the most fascinating and the fastest growing area of brain research today.

Secondly, prompted by the development of the transistor, the printed circuit board, and the moon exploration program, the electronic miniaturization revolution has advanced by leaps and bounds in the last decade or so. Circuit boards and transistors can now be produced which can be passed through the eye of a needle. Complex computers that were formerly as large as a room or even house have already been reduced to table top size, or less. Simpler computers can now be held in the hand, placed in the pocket, and purchased in a drugstore.

At the moment, it is difficult to predict where the electronic revolution as a whole is headed, but there have been some rather interesting specific, limited predictions made. For example, in March of 1977, Robert N. Noyce, chairman of the Intel Corporation, predicted that within 15 years semi-conductor chips would be produced with 2,000 times more memory storage elements or "bits" than today's chips have, at a cost equal to today's. He predicts chips capable of storing one million "bits" of information or 250 "logic" circuits, the equivalent of some of today's computers.[27]

Thirdly, there have been some rather astounding developments in computer research in the last few years. For example, computers have now been built that can understand as many as 1,000 human spoken words and thus can be controlled to at least some degree with the human voice.[28, 29] The progress to date falls short of a computer system such as "HAL" in the movie *2001,* but some are predicting advanced talking and listening computers within 10 to 15 years.[30] But even more dramatic than that, there is also research currently in progress whose goal is controlling computers with brain waves. The feasibility of direct brain-to-computer commands was demonstrated in 1976 at the Computer Science Department of the University of California at Los Angeles.[31] And, of course, the question of artificial intelligence in computers is still open. Although we admittedly do not yet have a computer which can improve and/or reproduce itself, we may be close to developing "thinking" computers. Today's computers can set up goals, make plans, consider hypotheses, recognize analogies, and "learn" tic-tac-toe, blackjack, chess, etc. Does that make them "intelligent" and able to think to at least some degree? The question could be argued, of course. But there are those who believe that the current gap between the computer and the brain will eventually be bridged—no matter how wide it might be considered to be today.[32, 33]

Could the three areas of development just discussed be blended with others discussed earlier in this chapter (e.g., artificial eye research) to eventually produce a man-computer chimera or mental bionic being? According to Rockefeller University psychologist and computer expert Adam Reed, that day could be closer than most people suspect.[34] At the 142nd annual American Association for the Advancement of Science meeting in Boston, in February of 1976, Reed suggested that mini-computers may someday be implanted in the human brain to aid the memory and provide users with "an almost infinite data capacity, together with reliability incomparably greater than the storage mechanisms of the natural brain."

The mini-computer could be considered to be an automatic brain booster, programmed to "read" and "speak" the electro-chemical language of the brain. One could store any information one wanted in it, and it could thus be a direct link to all of the world's available stored knowledge—with instant recall! Reed also believes one might be able to calculate the most complicated mathematical problems with split-second speed, contact other implanted computers instantaneously, and even become psychic in terms of reading the minds of those with implants. He admits there are at least five tremendously imposing problems to be solved before all of this is possible. Researchers will have to learn (1) how to build a complex mini-computer no bigger than a sugar cube (a 10-fold increase over current miniaturization), (2) how to get computer information into the brain, (3) how and where to hook the computer up to the relevant neurons or parts of the brain, (4) how to teach the computer or program it to speak the brain's language, and (5) how the brain stores, codes, and processes "the meaning of things." Reed, 30 years old, has said, "I don't know how long this will take, but we can expect it within our lifetimes."[34] Others have agreed, pointing out that the technological spadework has already been done.[35]

Going a step beyond Reed, one has to wonder if someday we might not produce "super scientists" by implanting mini-computers in the left hemisphere of the brain, "super humanists" by implanting them in the right hemisphere, or even "super human beings" by implanting interlinked computers in both?

If it becomes possible to keep a disembodied brain alive which contains an implanted mini-computer, another fantastic possibilty could open up—direct communication with that brain! What could it tell us? Arthur Clarke, British scientist and author, provided at least one answer when he wrote some years ago,

> Your present brain, totally imprisoned behind its walls of bone, communicates with the outer world and receives its impressions of it over the telephone wires you call the nervous system. These wires vary in length from a fraction of an inch to several feet. You would never know the difference if those "wires" were actually hundreds or thousands of miles long, or included radio links, and your brain never moved at all.[36]

Does the brain actually need the body? Might not the body in a sense, be a "drag" on the brain? Relieved of the many control functions associated with the body, might the brain achieve "new" heights of thinking, logic, and sensation as a result of its new experience of being isolated? Would "new" relationships and experiences with computers and machines be possible? Today we *control* computers, spaceships, TV sets, etc., connected by wires or radio links to these machines, tomorrow might not a disembodied brain *become* one? It seems almost impossible that the cyborg or disembodied brain that science fiction has described for so many years become reality. Or does it?

MENTAL MODIFICATION "INTANGIBLES"

Before we conclude our discussion of mental modification, there are several "loose ends" remaining that deserve at least passing mention. Actually they might better be termed "intangibles," for they involve areas of research whose futures are basically just not predictable at the moment. Any scientific research underway in each case is in its infancy and could either result in unexpected, tremendously-significant discoveries, or in a dead end!

The first intangible involves the fact that many experts believe a person genetically inherits half or more of his mental abilities, traits and instincts. If that's actually true, and if at least part of what a person is mentally is recorded in DNA and/or RNA, it could be possible to eventually use genetic engineering to "rewrite," edit, or "erase" at least part of that mental record! That intriguing possibility is discussed in greater depth in chapter 2 of Part II. The other intangibles that might be mentioned are biofeedback, hypnosis, ESP, telekenesis, levitation, pre-cognition, Kirlian photography, etc. If one, several, or all of these currently controversial abilities or phenomena could be proven to exist to *everyone's* satisfaction, with solid and convincing scientific evidence, the brain would be proven to have powers far in excess of those recognized today. It is difficult to predict where the extensive scientific research which this rather astounding "revelation" would prompt might lead.

IN CONCLUSION

There are still three final comments that should be made about brain research and neural science. First of all, in 20 or 30 years the man-in-the-street will probably realize what most space experts already suspect: beyond a certain point, space exploration will probably prove to a cruel hoax. As we explore the solar system beyond Mars, it seems highly likely we'll only find frozen, barren, lifeless, planetary chunks of death! And, with the nearest star unreachable, a realization will set in that we're marooned on our nine-planet island in the sky, and space will close as a frontier. But humanity needs frontiers, and will look for another one! Will we turn to neural science? We already are! Neural science is already at least one of the fastest growing areas of biomedical research today. A few years ago there were only hundreds of neuro-scientists. Today there are thousands!

Secondly, it could admittedly prove true, as some claim, that the brain and central nervous system will never be completely understood because of their extreme physical complexity on a micro-scale. However, it should also be realized that mankind already has some extremely powerful weapons in its arsenal for attacking such a problem—even on a molecular basis. The electron microscope, the computer, the use of chromotography and antibodies to differentiate between complex chemicals, and the use of radioisotopes to tag and follow the path of complex chemicals have already produced impressive progress in only the decade or so that they've been used.

Thirdly, it may also be true, as some claim, that, in addition to its physical complexity, the brain has a non-physical, mystical, spiritual component which will never be defined analytically. However, whether or not we ever *completely* understand the brain is perhaps irrelevant at the moment. As stated earlier in this chapter, it will not be necessary to completely understand the brain in order to begin to control it. And it seems likely that increasing understanding will produce increasing control.

Sir John Eccles, a leading brain researcher, summarized these three comments several years ago, when he said (footnote 10):

> Brain research is the ultimate problem facing man. . . . I predict that in three decades most of the greatest scientists in the world will be working on the brain. . . . Some people have speculated that, logically speaking, it may prove impossible for the brain to understand itself. I do not believe we are going to come up against a magical blank wall of this kind. But even if we should, our efforts to understand the brain can go on for hundreds of years without even being bothered about the

epistomological problem of whether the brain can, in the end, understand everything about itself.

Sir Charles Sherrington was a researcher of the nervous system in the early 1900's. The brain, Sir Charles decided at the age of 83, is an "enchanted loom where millions of flashing shuttles weave a dissolving pattern, always a meaningful pattern, though never an abiding one."

The brain, the "inner universe," is the next frontier!

CHAPTER 4

THE PROLONGMENT OF LIFE

> By the reproduction of cells, life thwarts time. Under the best circumstances the life span of individual cells is measured in days, weeks, months—at most decades; the slope of time is the declivity of aging. But time can be reversed, with 100 per cent profit to boot, by the reproduction of a cell.
>
> Each cell may begin its individual existence endowed with all of the potentialities of its parent and may annihilate its individual existence in the production of two cells that inherit those potentialities unaged and undiluted. The daughters of these daughters may do the same and so on to immortality.
>
> —Daniel Mazia, *How Cells Divide,* 1961

The dawning of modern medicine, drugs, surgical techniques, sanitary conditions, agriculture, etc., in the seventeenth century also ushered in an upturn in man's life-span. Life expectancy increases in the United States have been particularly dramatic in the last century or so, increasing from an average of about 41 years in 1850, to 48 years in 1900, to about 70 years in 1950. But the trend then leveled off, and life expectancy today is still about 70 years. Some believe the trend can be reinitiated and man can live even longer.

THE CONTROL OF DISEASE

One path to longer life-spans can be built with the increasing control and perhaps even eventual elimination of disease. Such control and elimination might be achieved in several ways. For example, as time goes on, an increasing number of diseased and malfunctioning organs will probably be replaceable with regenerated, transplanted, and mechanical organs. Computer diagnosis, laser surgery, improved chemotherapy, and new vaccines will also lead to greater control. Many researchers, for example, believe that cancer will be cured by 1985 with improved chemotherapy and/or new vaccines.

The development of genetic engineering techniques could also produce powerful new methods of controlling and eliminating disease. For example, the possible control of such genetically determined physical diseases as galactosemia has already been mentioned.[1] It may even be possible eventually to genetically correct certain forms of mental disease, if schizophrenia, mental retardation, etc., are actually caused by chemical deficiencies created by genetic defects, as many believe.

The development of genetic engineering techniques may also help solve the interferon problem. Interferon is a protein, discovered in 1957, which is manufactured at the ribosomes within a cell. Under normal conditions, virus infections release the genetic blueprint for interferon production. Within hours, interferon is manufactured and sent out to protect healthy cells. Interferon is thus the body's chief line of defense against virus infections and diseases. Obviously, injections of this protein, or stimulation of its natural production inside the body might provide immunization against all invading viruses. Work on exogenous interferon—that produced outside of the body and then injected — is proceeding slowly because of the problems involved in attempting to produce pure interferon in cultures. Stimulation of production inside the body has also been only partially successful. There is evidence that the production scheme suffers from fatigue if it is stimulated continually. Knowledge obtained in genetic engineering research could provide solutions to these problems and could thus help eliminate virus-caused infections and diseases.

Enzyme therapy may be yet another route to controlling and even curing various diseases. There are up to 1,000 enzymes in every cell in the body. They are proteins which take orders from the RNA present, and then in turn dictate and control the reactions that occur in the cell. They also prevent the accumulation of intermediate products which might favor harmful alternative pathways. Enzymes function by acting as catalysts. They are capable of increasing the rates of cellular reactions manyfold under the relatively mild temperature and acidity conditions that exist within the cells. The enzyme content of cells and organs varies if there is disease present. Many believe that various metabolic and genetic disorders (such as Tay-Sachs disease), that result in a deficiency of certain enzymes, might be controlled or even cured by additions of those enzymes, once they can be produced synthetically. The control of diabetes has already shown it can be done.

It is difficult to predict what might eventually be accomplished through enzyme therapy. For example, in 1969, Dr. Arthur Kornberg of the Stanford School of Medicine announced the discovery of DNA polymerase, an enzyme that apparently can repair damaged chromosomes.[2] According to Kornberg, many diseases result from chromosomal damage and the resultant failure of DNA polymerase to do its repair job properly for some reason. He has predicted that this enzyme, and perhaps other undiscovered repair enzymes as well, might be placed in "sick cells" by using "friendly" carrier viruses and the transduction technique. Kornberg has proposed a massive research search for "friendly viruses" with what he calls "fortunate characteristics," although he admits such a search will be expensive.

With the development of techniques such as those just discussed, and perhaps others as well, it seems safe to say that disease will be controlled and perhaps even largely eliminated by the year 2000 and possibly even before. The control and elimination of disease will bring increases in life expectancy, just as they have in the past.

FREEZING TECHNIQUES

A book written in 1964 by Robert Etlinger, titled *The Prospects of Immortality,* ushered in a new approach to the prolongment of life. Conceding the fact that everyone eventually dies, Etlinger and his followers advocate freezing an individual to −321° Fahrenheit (with liquid nitrogen) as soon as possible after death, keeping the individual in this state until what killed him can be corrected or cured, and then attempting reanimation. The slogan of people advocating this technique has only semi-humorously been, "Freeze, wait, reanimate!"

It is difficult to determine how many people there are in the United States and the rest of the world already "on ice," for statistics are not readily available. The most publicized freezing to date has probably been that of Dr. James Bedford, a retired psychology professor from Glendale, California. Bedford died of lung cancer on January 12, 1967. He was packed in ordinary ice as soon as possible after death, later in dry ice, and two weeks after that was moved to Phoenix, Arizona where he was placed in a capsule cooled with liquid nitrogen to −321° F. The plan is to thaw and reanimate Bedford when lung cancer can be more effectively cured, corrected, or repaired.

Will such a freezing technique actually work? Most of the scientists who study the effects of low temperature on living systems (cryobiologists) say "no" —not in the foreseeable future and probably not ever. In support of such statements they point out two very important factors that they believe work against such a scheme. First of all, water has the unusual property of expanding when it freezes. The freezing of water will thus damage radiators, sidewalks, highways, etc. The same damage should occur in cytoplasm, which is chiefly water—and it does! Such damage in living tissue might be expected to be irreversible—and it is! Secondly, every living cell requires oxygen in order for cell metabolism to take place. A human being can remain in a state in which breathing, circulation, and the heart have stopped (clinical death) for only about four minutes, under normal conditions. After four minutes extensive irreversible cell damage occurs due to a lack of oxygen (especially in the brain), and a state known as biological death results. Revival from clinical death is fairly commonplace, but no one has ever been revived from a state of biological death! Obviously, anyone frozen after "death" would be in a state of biological death. It's doubtful whether freezing could be initiated within the four minute period of clinical death. If it were, the question of murder might arise because of the possibility of revival in this period.

The two arguments just raised in opposition to the freezing techniques that some are advocating seem almost overwhelming at first glance. However, those who favor freezing believe there are possible loopholes in them. First of all, they point out that the cellular damage caused by freezing can be minimized today and will, perhaps, even be eliminated completely in the future through the use of antifreeze, much as in an automobile radiator. The use of glycerine and dimethyl sulfoxide (DMSO) in this respect has already made possible the freezing of blood, marrow, cell cultures, and sperm for up to ten years. In 1972, several researchers at the Oak Ridge National Laboratory announced that they had developed a technique for producing live, baby mice from mouse embryos which had been frozen for as long as eight days in dry ice or liquid nitrogen.[3 4] The frozen embryos were then slowly thawed, were allowed to grow in a dish for a few hours, and were surgically implanted in "foster" mice mothers. DMSO was used as the antifreeze in the experiments. Similar research with frozen calf embryos was apparently less successful.[5]

The ability to freeze blood, cell cultures, sperm, and embryos is obviously a step forward for cryobiology (and artificial inovulation). But how far forward? The ability to freeze organs for extended periods of time and establish frozen organ banks would clearly be a much more impressive step forward. Before 1976, however, cyrobiologists had been generally unsuccessful in attempts to do it.[6] In late 1976, another research team at the Oak Ridge National Laboratory reported the development of a method for slowly freezing pancreases from rat fetuses for periods of days to weeks.[7] After thawing, the pancreases supposedly synthesized 80 to 100 percent as much protein as did fresh, unfrozen organs. The researchers hope the new technique will lead to the successful freezing of other mammalian organs and perhaps eventually to frozen organ banks.

But even if the freezing point barrier for freezing any organ is overcome, and frozen organ banks become commonplace, the barrier for freezing whole bodies could remain. Those who advocate freezing believe that we would be well on the way to conquering the body barrier by conquering the organ barrier. Most scientists are more skeptical.

Freezing advocates also believe that ways may be found to alleviate or even eliminate the oxygen-deprivation, cellular damage that results in biological death. They rest their hopes on the fact that bodily processes slow down and oxygen consumption decreases with decreasing temperature. It is true that, for these very reasons, humans are now being routinely cooled to 50–70 degrees, in open heart and brain operations.[8] And it is also true that there have been instances of people found frozen to such temperatures who have been revived. However, there is cellular damage which takes place with cooling, and, at present, cooling below 50 degrees can be fatal. And thus there is and has been only very limited research conducted in attempts to get past this barrier, for it has been extremely difficult to obtain volunteers for such experiments.

In 1955, one of the few known tests conducted to determine the lowest temperature a human being could survive was conducted in the United States.[9] The volunteer subject was a woman suffering from terminal cancer. Her body temperature was lowered to 48.2 degrees F. At this temperature, her respiration and heartbeat stopped, and she lay in a state of clinical death for 45 minutes. It took 15 minutes of rewarming her body before her heart started beating again. After 10 hours, she regained consciousness, with no apparent after-effects, but died of the cancer 38 days later.

It's admittedly a giant leap from reviving people intentionally cooled as low as 50–70 degrees, for surgery or research purposes, or unintentionally by lakes or snowbanks, to reviving people cooled for years to −321°F. But it is at least a small step in that direction.

Could a person frozen for 10, 50, or 100 or more years actually be reanimated some day from a state of biological death? The technical obstacles involved in overcoming the associated problems of freezing damage and oxygen-deprivation seem almost overwhelming. But research directed toward these problems is underway. The research is in its infancy, but it should be remembered that the flight of the Wright Brothers led to the development of the Boeing 747 and the SST within 65 years!

It should also be pointed out that, although the characteristics of "life" can be defined today, our actual understanding of what "life" *is* is very meager. To be able to define or describe something is not necessarily to understand it! Is a seed "alive" before it's planted and begins to grow? At what point does an embryo or a fetus begin "living"?

According to the February, 1972 issue of *Soviet Union,* a magazine printed in the U.S.S.R., a Soviet geochemist had been able to "revive" some bacterial organisms found imbedded in a sample of potassium ore that was 250 million years old! In December of 1973, two researchers at the Darwin Research Institute claimed to have revived some bacteria found in rock samples bored from a depth of 420 feet in Antarctica. By the usual definitions of the word "life," the bacteria in this second instance had been "dead" for 10,000 to 1,000,000 years![10] A type of Tardigrade was discovered in 1975 which had been able to exist for more than 100 years without water, oxygen, or heat—in a dehydrated state, which, by most definitions would be called "death." And yet, when it was moistened, it immediately sprang back to "life"! Several researchers have coined the word *cryptobiosis* to describe the phenomenon.[11]

Are there undiscovered, perhaps even unexpected, things yet to be found out about "life" and "death"? Are these two states really separated by a wall, or are they actually intertwined in some way? Are they simply manifestations of some third, currently unsuspected state? It could very well be that the freezing advocates are overly-optimistic, but it could equally be true that the critics are overly-pessimistic. Perhaps the prospect of revival from death should be placed in the area of possibility rather than impossibility. After all, at one time not so long ago, it would have been considered to be impossible either to restore life to dogs who had stopped breathing for two hours, Tardigrades "dead" for a century, or to bacteria that had been buried for thousands to millions of years. What other surprises might lie ahead?

CHEMICALLY-INDUCED HIBERNATION AND SUSPENDED ANIMATION

It seems reasonable to believe that the prolongment of life may be accomplished sooner through induced hibernation (a slowing of the body processes) than through the use of freezing techniques (a complete cessation of the body processes), since the technical problems involved seem less formidable.

Hibernation is not simply a deep sleep. It is a state of suspended animation that profoundly slows every bodily process and lowers body temperature. For example, a hibernating bat's temperature may approach freezing. A hibernating marmot takes only one short breath every five minutes. A hibernating ground squirrel's heart beats weakly three times a minute compared to a normal 360 beats. It can be shown rather definitely that slowed body processes and lowered temperature result in a prolongment of life. For example, some hibernating bats live 20 times longer than non-hibernating bats of the same weight.

Oddly enough, research on hibernation has only been underway for a decade or so, probably explaining the fact that the hibernation mechanism is not yet understood. However, research is proceeding on several fronts, and some rudimentary progress is being made.

Dr. Vojin Popovic, Professor of Physiology at Emory University, is making a rather direct attack on the problem. In late 1974, Popovic announced that, after studying various animals in a state of hibernation, he had been able to induce temporary clinical "death" or suspended animation in 12 dogs, and then revive them with no signs of permanent damage.[12] The dogs were slowly cooled by pumping their blood through a heart-lung machine containing an ice-water device, while packing the dogs externally in ice. At 40° F., the blood circulation was discontinued and clinical death began. There was no electrical activity in the heart and brain, and no blood circulation or blood pressure for two hours—a record for induced clinical death in non-hibernating animals. Later circulation and blood pressure were gradually restored, and the heart and brain returned to normal. Thirty minutes after reaching normal temperature, the dogs were able to walk and drink water, and a few hours later were able to eat. According to Popovic, "If we can do this with dogs, it can also be done with man." He believes that such research may some day allow surgeons to operate on patients for hours without fear that lack of circulation would do crippling damage, and that space travelers may also be kept in suspended animation for long voyages. Popovic has also said that, "If man could hibernate, he could possibly live to about 1,400 years." (See footnote 12.)

Other researchers are taking the approach that it might ultimately be far more practical (although not necessarily easier) to induce hibernation or suspended animation with chemical rather than physical techniques. And thus there is work underway directed toward isolating, characterizing, and synthesizing the chemical(s) in the brain and/or bloodstream that may cause hibernation in animals. In early 1974, two researchers at the Stritch School of Medicine in Maywood, Illinois, announced that they had found that an injection of serum from the blood of a hibernating squirrel could also put a squirrel that was awake into a state of hibernation.[13] Unfortunately, in this and similar experiments done before and since, the specific chemicals responsible for the hibernation have not been identified as yet. Even if and when they are, there is no assurance, of course, that they will work on humans.

There may also be three other possible sources of help in solving the hibernation riddle. First of all, there is the theory of some that the whole hibernation process may be under genetic control. If that is true, it might be possible to study and perhaps even induce hibernation with genetic engineering tecniques. Secondly, it is quite possible, many say, that hibernation, sleep, and anesthesia may all share some common chemical and molecular basis. If that is true, research currently underway on sleep and hibernation could also provide some meaningful or even valuable clues about hibernation.

Unfortunately, although a tremendous amount of research has been done on sleep, there has been very little progress made in determining what starts and stops it, and how people move from one phase of sleep to another. The brainstem, thalamus, basal forebrain, and hippocampus have been identified as areas of the brain that influence sleep, but which neurons and nerve chemicals in these areas are involved in the control?

There has been some progress recently in answering these questions. In 1974, a Swiss research team reported the discovery, isolation, and identification of "sleep peptide delta," a chemical found in the bloodstream of sleeping rabbits.[14] In 1975, several researchers at the Harvard Medical School announced they had found that two different neurons act as "sleep switches" in the waking sleep cycle.[15] The nature and function of those neurons is currently under investigation. In June of 1976, a researcher at the University of Texas Medical Branch at Galveston reported that he had been able to alter the sleep and behavior patterns of monkeys by injecting an enzyme into their bloodstreams that depleted some of the amino acids present.[16] He thus believes that certain amino acids in the bloodstream can influence both sleep and behavior. As further research results on sleep are announced, some light may also be shed on hibernation, since both sleep and hibernation are forms of regular physiological narcosis.

Despite vast amounts of research since the advent of clinical anesthesia more than a century ago, the mechanism of action of general anesthetics also unfortunately still remains a scientific challenge. There have been a number of theories proposed, but the exact mechanism by which anesthetics cause physiological narcosis still remains to be determined. This should hardly be surprising in view of what has been said previously about sleep, and since even the normal state of consciousness is poorly understood. The search for understanding about anesthetics is beginning to move more and more to the molecular level, and this is encouraging, for that is where the specific answers are. As those answers come, they too may shed some light on hibernation.

A group of researchers at Michigan State University have probably done the most significant, and the most unique, work to date on the chemical inducement of hibernation and suspended animation. Their work seems to carry a promise or feeling of ultimate success, although admittedly only time will tell whether that success is actually achieved. In March of 1974, the group publicly announced that they were accelerating research on a drug which they believe may ultimately triple man's life span. The drug, as of yet unidentified in press releases, would apparently lower one's body temperature, and thus achieve the desired life-span increases promised by hibernation and suspended animation—*but without the associated narcosis!* Thus far, tests conducted on fruit flies have been quite dramatic. By depressing the temperature-control center in their brain, the aging process in the fruit flies has apparently been slowed down. One group of about 400 flies, with their temperature controlled at about 91° F., lived an average of 10 days. Three other groups controlled at about 88, 81, and 77 degrees, lived averages of 20, 50, and 70 days respectively! Research has recently been initiated on mice, because they have a temperature control system much like humans.

Dr. Barnett Rosenberg, co-leader of the research team, believes that positive results with mice could lead to the drug being tested on humans within 10 years. He has theorized that lowering the body temperature to about 95 and 87 degrees could increase the life span to about 90 and 200 years respectively. He concedes that possible use of the drug may be delayed years by serious questions about possible physiological and societal side effects, but in a press release he has been quoted as saying that, "It is time the public became aware of the implications of such preliminary studies of aging."

Can man hibernate someday, or be placed in a state of suspended animation? Could he reap the benefits of such a state without having to experience the associated narcosis? In his 1963 book *Profiles of the Future,* Arthur C. Clarke picks 2050 as the date by which such techniques may have been perfected.[17] However, research conducted since his prediction in 1963, currently underway, and projected in this area, would seem to indicate that success could come well before that, possibly by 2000. Some now believe even sooner!

CONTROL OF THE AGING PROCESS

Three methods have now been discussed as possible means of prolonging life. However, it's difficult to envision the development of freezing techniques before 2000—if then. The development of hibernation and suspended animation techniques appears to be possible, but doubtful, much before 2000. Although the control and elimination of all or most major diseases may be accomplished well before 2000, most experts believe that man may still live only an average of about 15 years longer as a result. For example, the person who would no longer die of a coronary at 90 would be so weakened by old age at 105 that any trivial cause could then cause death. Realizing the above facts, many researchers believe that the best way of providing significant life prolongment the soonest may be through actually solving the aging puzzle. Thus it is that aging research has been receiving more and more attention in the last decade or so.

Aging is one of those words we think we can define, until we try. Actually, about all even science can do is describe what everyone knows *happens* as a person ages. The skin wrinkles, the hair turns grey, the hands grow gnarled, the bones become brittle, the eyes no longer focus properly, body bulk declines, and eventually, there's death. The whole process currently takes 70.4 years on the average in the United States—66.6 years for men and 74.2 years for women. But what determines these averages? What actually causes the physiological changes in the body that we call aging? These are questions which science cannot answer at the present time. The questions are particularly perplexing because theoretically we shouldn't age!

Each human being starts out as a tiny, one-celled, fertilized ovum, which begins dividing soon after it is formed. As mentioned earlier, an average adult is comprised of the approximately 100 trillion cells which result. The cells are essentially duplicates, but because of differentiation, there are differences in how they function. Perhaps the biggest difference between the cells, however, is that some are permanent (non-dividing) and others non-permanent (capable of division). Muscles, nerves, the heart, brain, kidneys, lungs, and some other organs consist of permanent cells. The skin, blood, hair, fingernails, and some organs such as the liver consist of non-permanent cells. Non-permanent cells have a limited lifetime, as short as two or three days in the case of the skin and blood. As these cells die, the body replaces them—as many as three billion per minute! However, theoretically, the replacements should be exact duplicates of the originals. In addition, permanent cells should theoretically not die or stop functioning. And so, overall, theoretically, we shouldn't age. Many researchers believe that if a person could remain as healthy and "young" as he was at 15–20 years old, he should be able to live for 800 years or more.

Man does age, however, with the process accelerating markedly at about the age of 40. Non-permanent cells replaced by the body gradually become inferior to those originally present and the body's repair processes gradually slow down. But even more important, permanent cells gradually die. The body loses up to 40% of its permanent cells by the age of 75, as compared to those present at 20.[18] Some of the bulk lost is replaced by fluid or connective tissue, but this does not prevent the somewhat amazing effects of the loss. By the age of 75, muscle weight and thus strength may fall by a third as compared to that at 20; there is a shrinkage in height because of a loss of collagen (fibrous protein) between the vertebrae of the spinal column; the lungs may be only half as efficient; the amount of blood pumped in the body may drop as much as a third; only half of the kidney cells may remain; a fourth of the nerve neurons may be gone, resulting in diminished reflexes and reaction times; sight, touch, taste, and hearing markedly diminish; even the number of taste buds may diminish, often by as much as 60%. The most serious effect of aging, however, was probably described by Shakespeare in *Much Ado About Nothing* when he said, "When the age is in, the wit is out." Between the age of 20 and 75, the weight of the brain may drop from an average of 3 pounds to as little as 2¼ pounds, resulting all too often in senility.

Interestingly enough, different parts of the body may also age at different rates. For example, a 40-year-old man may have 35-year-old lungs and a 45-year-old heart. Or, a person of 70 may be in good physical condition and senile—or just the reverse. It's also interesting to note that practically no one dies directly of old age. Most people die of the weakness and loss of resistance that old age causes. In old age there is increased vulnerability to the infections and diseases which the body might have survived when younger. At 90, even a common cold is a mortal enemy!

It's quite obvious that the answer to the aging puzzle is inside the cell. And since the cell is a chemical factory, the aging problem is obviously a chemical problem. The answer will, therefore, be a chemical answer. But that's about all that the scientist can currently say with any certainty, for aging research is only in its infancy. Such research only began about 1950, and has really only grown to significant proportions in the last decade or so. About all that has been accomplished so far is the formulation of theories. In fact, as many as 120 theories have been advanced as to the causes of aging! The overall picture is so confusing, that as researchers sift through the theories, it's often easier to attempt to discard unlikely theories than to pick those that seem more plausible.[18, 19, 20, 21]

There are at least six reasons why it will be difficult to pinpoint the exact cause(s) of aging. First of all, it's now beginning to appear that several theories may be correct. There is a growing feeling that there may be several interrelated causes of aging. Secondly, the answer is probably inside the cell, but the size of the cell, chromosomes, and genes, as discussed previously, makes aging research extremely difficult. This leads to problems three and four. How do you separate and then analyze the extremely small quantities of chemicals found in the cell? The complexity of these chemicals is also a problem in analysis. A typical enzyme, hormone, or protein can contain hundreds, thousands, or even millions of atoms. Fifthly, the thousands of chemicals present in the cell all interact to form a harmoniously operating, but extremely complex chemical factory. Understanding the role of just one chemical in the cell can be much like trying to determine what a jigsaw puzzle looks like by just looking at one piece. And lastly, it is difficult to conduct aging research because of the time required. In studying aging or possible remedies for aging, it would be best to study a large number of subjects, originally 20–30 years old, over a period of 40–50 years. But because this is not possible, short-term studies of about 5 years or so are usually attempted. Such studies are usually plagued by the usual inconsistencies and inconclusiveness encountered when any long-range, slowly changing phenomenon is studied over a short period of time. In a recent issue (March-April, 1976, page 164) of the *American Scientist,* there is a cartoon that does an excellent job of summarizing the testing dilemma in aging research. Two scientists are discussing a new anti-aging drug one of them has discovered, prompting the other to say, "It may very well bring about immortality, but it will take forever to test it!"

Despite the great number of theories about aging now available, many have been discarded, or are no longer seriously considered. Some are thus automatically left in a more-favored position. Briefly outlining ten of the "more favored" theories may help illustrate the current confusion, complexity, and perplexity connected with the aging puzzle.

Interestingly enough, six of the ten do not require or directly involve DNA in their explanations. For example, the "biological insult" theory simply says that after a series of varied and harsh bodily "insults" (poor diet, insufficient exercise, smoking, etc.), the body simply wears out. The "diet" theory zeros in on diet as being the major culprit among the "insults."[21, 22] Dr. Alex Comfort of University College in London is one of the leading advocates of the diet theory. He believes overeating and obesity go hand in hand with aging and shortened life-span, but that undereating and underweight may actually have a lengthening effect of life-span.

The hormone theory says aging accelerates in the 40's of a person's life because those are the years when the body gets increasingly "out of whack" due to an accelerating decline in hormone secretion. (See footnote 21.) (Note: hormones are chemicals secreted by various glands in the body, that help regulate bodily functions and cellular activity.) Unfortunately, there's a flip side to this theory. Does decreased hormone secretion cause aging? Or, does aging cause decreased hormone production?

The "trace metals" theory says that, over long periods of time, an increasing concentration of metals such as iron, copper, calcium, etc., cause cellular aging in some as yet unknown way.[23] Dr. Hans Selye, a noted researcher at the University of Chicago, is a leading advocate of calcium stress theory, which zeros in on calcium as the leading aging culprit among trace metals.[24] Selye believes increasing accumulations of calcium in the body, prompted by sustained physical or emotional stress, cause aging.

Dr. Dale Harmon, of the University of Nebraska, is one of the leading advocates of the "free radical" theory. (See footnotes 19, 21). Free radicals are extremely reactive molecular fragments that result when a chemical bond in a stable molecule is split or cleaved by any of a number of causes. Although free radicals are very short-lived, Harmon believes they "attack" the chemical composition of normal cells, causing cellular deterioration and subsequent aging. The free radical theory is one of the more recent aging theories, but, perhaps one of the more promising.

There are three current aging theories that directly involve DNA in their explanations. The "built-in death" theory is probably the more intriguing of the three. It says that just as there is a blueprint or code of life in DNA, there may also be a blueprint or code for aging and death.[25] In other words, living things may have a clock or timetable for aging and death built into what might be called aging and death genes.

The cell damage theory says the DNA in a cell gradually becomes "garbled," producing gradually decreasing cell efficiency (aging) and, eventually, cell death. There are variations to the theory.[26] Different groups of researchers believe the "garbling" may be the result of mutation, accumulated errors in the self-replication for some reason, changes in the expression of certain genes due to a breakdown in gene repair, or changes in the expression of certain genes brought about by aging genes. The "garbling" may also result in the increasing solid, floating, pigment-like material which accumulates in cells as they age. The "clinker" theory says these "clinkers" or "metabolic ash" gradually clog or choke a cell, disrupt or short-circuit it, and thus cause aging.

The tenth theory we will discuss, the "cross-linking" theory, is perhaps the best known. It says that aging is caused by the progressive intermolecular cross-linking of proteins and/or nucleic acids (DNA and RNA). Some researchers believe that cross-linking may cause the progressive loss of flexibility and lessened function that tissue and organs undergo with aging. They also believe that cross-linking may prevent some protein molecules from being metabolized by the body's enzymes. As a result, these proteins collect in the cells, may clog them, and ultimately destroy them. Dr. Johan Bjorksten, Director of the Bjorksten Research Foundation near Madison, Wisconsin, is probably the leading advocate of the cross-linking theory.[27] Bjorksten compares the cross-linking of proteins to a big room, where thousands of people are working, and each day some person comes in and handcuffs a pair of them. Obviously, the work will gradually slow down and eventually come to a stop. Alternatively, one might think of the body as being a bowl of spaghetti, the spaghetti representing the long, polymeric protein molecules of which the body is essentially comprised. When young, the body is like a bowl of hot spaghetti, with the strands independent and flexible. When older, the body is like a bowl of cold spaghetti, with the strands stuck together and inflexible!

One of the advantages of the cross-linking theory is that it can be used to explain most of the others. It does not explain the "built-in death" theory. However, the "garbling" in the cell damage theory might be explained simply as the increased cross-linking of DNA. The "clinkers" that accumulate in cells could be explained as cross-linked "metabolic ashes." A decrease in the secretion of hormones could be explained as being caused by cross-linking in glands and/or the hypothalamus. The biological insult theory can be explained in terms of the "insults" (radiation, smoking, etc.) causing cross-linking. The diet theory can be explained in terms of overeating causing the accumulation of cross-linking agents in greater quantities than the body can metabolize or eliminate. In the metals theory, one can simply point out that calcium and other divalent metals are cross-linking agents. The free radical theory can be explained in terms of free radicals causing cross-linking as one of their major effects.

Despite its theoretical attractiveness, the cross-linking theory is currently "stuck on center." It seems to be neither picking up or losing strength and support. Everyone agrees that cross-linking occurs in the body, but no one has demonstrated to everyone's satisfaction that cross-linking actually causes aging. Analysis problems have been quite severe in this respect.

Undaunted, Bjorksten goes on to theorize that normal enzymes are unable to break down the cross-linked protein masses that accumulate in and around living cells, and that aging, sometimes accompanied by disease, ensues. He reasons that such enzymes must exist because the unwanted protein accumulations do break down naturally after death, and this must be through enzyme action. He also points out that soil bacteria and molds must contain such enzymes, for otherwise the earth would be covered with deposits of cross-linked protein. Since the late 1960's, Bjorksten and co-workers claim to have actually discovered and isolated seven previously unknown enzymes capable of dissolving the unwanted proteins. Bacillus cereus was the first of the seven discovered, and thus it has been more extensively investigated than the others.[28] Experiments wth Bacillus cereus in Bjorksten's laboratory, on animals, have supposedly been very promising. The enzyme is now being tested in various other laboratories around the world, under different conditions, to further determine its effectiveness and applicability.

Even if Bacillus cereus turns out to be a failure, and all of the rest of the six follow suit, Bjorksten believes that a 10 to 30 year period of intensive search for such an enzyme or agent will be successful and that a way will be found to actually re-mobilize aged protein. Of course, the agent found must be non-toxic and capable of reaching the sites of immobilization—two rather large orders. But Bjorksten (and others) believes it can be done.

IN CONCLUSION

There is a growing number of scientists working in the aging field. According to one estimate, the number of United States researchers in this area has increased about 20% in the past 10 years. There is increasing money available for research on aging and also mounting public awareness of the desirability of increasing research. People are beginning to realize that cancer attacks about 1% of us, heart disease about 10% of us, etc., but aging attacks all of us—100%! In its February 1972 issue, *Industrial Research* disclosed the results of a questionnaire on aging research answered by about 1,850 scientists and engineers. 82% of those polled believed such research should be expanded. 65% believed there would be significant advances in the near future, although 61% felt that less than a 10-year extension of the life span could be accomplished by the year 2,000. 85% felt that expanded life spans would provide significant side benefits to society-at-large.

If an anti-aging agent or technique is found, what can we expect in the way of results? Probably, at best, a regression of aging which is not reversible and a maintenance of bodily resistance at the 15–20 year old level. However, this would be the ideal, and regression may be more difficult than prevention or a state of remission. Regression may also be more and more difficult the older the person is undergoing the treatment. What might the ultimate life-span extension be? As stated earlier in the chapter, if a person could remain as healthy and "young" as he was at 15–20, a life-span of 800 years or might be possible!

Throughout the ages men have spent hours, months, years, and even lifetimes seeking ways to prolong their lives and perhaps even achieve immortality. The keys to longevity have been thought to be everything from witches' incantations to vitamin E. On March 3, 1512, Ponce de Leon set out for the new world, searching for a rather unique key. He was looking for a fountain of youth, supposedly hidden somewhere on the island of Bimini in the Bahamas, that legend said rejuvenated the old and strengthened the weary. Ponce de Leon died nine years later—still looking! Another expedition left for uncharted waters in the 1950's, also looking for the fountain of youth. But this time it is powered by science, technology, and medicine, rather than sails, and seems to be on course, rather than drifting.

CHAPTER 5

THE CREATION OF LIFE

> . . . I felt slightly queasy when at lunch Francis [Crick] winged into the Eagle to tell everyone within hearing distance that we had found the secret of life.
>
> —James D. Watson, *The Double Helix*

THE CREATION OF LIFE

Man has always had a driving, compulsive desire to explain his origin. Before the time of the Greeks, primitive and ancient peoples had rather simple accounts that might be termed creation stories. The stories were usually religious in nature, ascribing the origin of life to a divine being(s). In the sixth century B.C., a Greek philosopher by the name of Anaximander proposed a theory not too different from the theory of evolution suggested by Charles Darwin in 1859. Anaximander believed that higher living organisms developed from lower types and that at one time man was a fish. He also believed that gradual evolution was prompted or stimulated in nature as water was evaporated by the sun. Later in the fifth century B.C., Democritus, Leucippus, and other Greek philosophers reaffirmed these beliefs. They also added the thought that life had originally emerged from a primeval slime. Plato strongly ridiculed all of these ideas, thus detracting from their popularity. However, Aristotle, a pupil of Plato, also believed that life arose *spontaneously* from non-living matter, transformed by a "vital force," in a moist environment generated by evaporation caused by the sun. At this point, the ideas regained their popularity and became known as abiogenesis or the theory of spontaneous generation.

Many biblical scholars now believe that parts of the book of Genesis were actually written in the fifth or sixth century B.C. However, although Genesis was later widely accepted in the western world, Aristotle's works were "lost" and remained largely unknown until about the beginning of the twelfth century A.D., when they were "rediscovered" during the Crusades and the "scholasticism" movement. An interesting thing then happened. Although it was an age of the literal interpretation of the Bible, the two viewpoints were merged to some degree. Many scholars soon came to accept the idea that God and the "vital force" could be considered to be one and the same cause for the origin of life on the earth, and the biggest obstacle to merger was thus removed. Abiogenesis also seemed to have a certain logic or "ring of truth" to it which was also quite apparent. For example, everyone could see that decaying meat did indeed bring forth maggots, garbage spread rats, ants sprang from honey, etc., all spontaneously, just as Aristotle had said!

And so abiogenesis was quickly and rather solidly planted in the minds of many scholars and remained there for centuries to come. Medieval alchemists came to dream of spontaneously creating life in a retort, as well as changing the base metals into gold and silver. However, it wasn't until the late 1700's that Mary Shelley foresaw what might be termed the first truly scientific spontaneous creation of a living being. The scientist in her novel was called Frankenstein, although the word has become synonymous with the monster he produced.

In the early 1700's, with the discovery of the microscope and the refinement of various scientific methods, abiogenesis started to come under verbal attack. In the early 1800's, Redi, Leeuwenhoek, Spallanzani, and other scientists performed experiments which rather strongly indicated that living organisms cannot be spontaneously formed from formless solutions and infusions. In other words, they rather convincingly showed that life could only develop from life. In 1828, a German chemist by the name of Frederich Wohler also dealt a severe blow to the doctrine of "vitalism." Wohler was able to produce an organic chemical (life consists of organic chemicals) from an inorganic chemical for the first time. Specifically, Wohler prepared urea (the chief component of urine) from ammonium isocyanate. He thus proved that organic compounds do not contain some magical "vitalistic" force which differentiates them from inorganic "dead" chemicals.

In 1859, Charles Darwin published his *Origin of the Species,* in which he said all living things, including man, had arisen through an evolutionary process. He even went so far as to propose a mechanism, and the words "survival-of-the-fittest" soon became almost a household phrase. At about the same time, A. R. Wallace proposed a similar theory. Neither Darwin or Wallace had much to say about the actual origin of life, although there are indications they both believed it could have arisen abiogenically.

In 1859, Louis Pasteur and others were well on the way to conclusively proving experimentally that the spontaneous generation of life was impossible.

The job was completed about 1865 and the spontaneous generation theory was in shreds. Many believed that Genesis had also been attacked and a storm was thus also unleashed in the religious world.

In the last few decades, chemists, biochemists, biologists, geologists, etc., have made great strides in answering many of the specific questions about life that Darwin, Wallace, Pasteur, and others raised. Chemically speaking, we now know that the basic materials of living matter are quite simple. Carbon (C), hydrogen (H), oxygen (O), and nitrogen (N) constitute 99% of living things. All are gases in the elemental form, except carbon, but this, too, when combined with oxygen, forms a gas, carbon dioxide, which is very important in life processes. Add eight more elements (calcium, sulfur, phosphorous, silicon, potassium, magnesium, iron, and sodium), and 99.98% of living matter is accounted for.

In living organisms, carbon, hydrogen, oxygen, and nitrogen combine to form four main categories of materials—carbohydrates, fats, proteins, and nucleic acids. The fats are simplest, each molecule consisting of three fatty acid molecules joined to a molecule of glycerine. Carbohydrates are chains of sugar units, and their function, together with fats, is to furnish fuel for energy for the living organism.

Nucleic acids (DNA and RNA) are more complex. They are very large molecules made up of aggregates of four smaller molecules (known as nucleotides) linked together in many possible sequences. Nucleic acids are believed to be responsible for the propagation of the species, as discussed earlier in chapter 2.

The most complicated components of life are proteins, built from about 20 amino acid units, hundreds and thousands of which are strung end to end in an almost infinite variety of sequences. No two living species of organisms contain the same proteins. This fact was also discussed earlier in chapter 2.

The problem of explaining the origin of life, at least chemically speaking, is then twofold. First of all, how did fats, carbohydrates, nucleic acids, and proteins originate on the primitive earth? And secondly, how or why did these units become combined or assembled in the right or proper manner to form life?

The foundation for attempting to answer the first question was laid in 1924 by two scientists, A. I. Oparin and J. B. S. Haldane. Each independently published books paralleling the other's. The two books leant credence to each other, and together brought the ideas expressed more effectively before the scientific world. Oparin and Haldane both suggested that the harsh conditions present on the early earth resulted in a primitive or primordial atmosphere comprised of methane (CH_4), ammonia (NH_3), hydrogen (H_2), and water vapor (H_2O). Scientists have since argued about the exact nature of the composition. However, regardless of what it was, no one now seems to disagree that the components present formed the first organic molecules through chemical reactions intiated by lightning and/or ultraviolet radiation from the sun. These more complex molecules then settled into the oceans and pools of water on land. In some manner, the molecules then gradually coalesced into amino acids, nucleic acids, etc., and into clumps of such molecules now called coacervates. The coacervates and molecules gradually changed, combined, and evolved, finally forming the first living organism. Some scientists believe it was a virus or virus-like material, although others question whether such a material is "alive." It's also debatable whether the first "life" arose in the oceans or in a pool of water on land. Wherever it was, the water involved is often referred to as the primordial "broth" or "soup."

Early, simple "life," upon prolonged evolution, ultimately developed into single-celled and then multicellular systems such as algae. There's no question about systems such as these being "alive." Further, prolonged evolution and interaction with the environment probably resulted in the development of metabolism. Early organisms probably began fermenting organic molecules, discharging carbon dioxide as a waste product. Then gradually they learned to use the carbon dioxide, releasing oxygen as a waste product. Specialization in one type of process or the other resulted in the establishment of carbon dioxide and oxygen cycles, wherein ultimately plants and animals used each other's waste products. Respiration and photosynthesis were thus established. Some plants and animals then moved from water to the land. Evolution continued, producing the increasing complexity of life with which we are all familiar. The time scale believed to be involved in all of these chemical changes or steps is shown in the following table.

THE HISTORY OF LIFE ON EARTH

From (B.C.)	*Historical Epoch*	*To (B.C.)*
4,600,000,000	Formation of proteins, nucleic acids, etc.	3,200,000,000
3,200,000,000	Development of viruses and living cells	2,500,000,000
2,500,000,000	Development of photosynthesis	1,500,000,000
1,500,000,000	Development of respiration. Life comes to surface of the sea.	650,000,000
650,000,000	Development of multicellular organism. Life moves to land.	230,000,000
230,000,000	Dinosaurs and plant life	70,000,000
70,000,000	Mammals	2,000,000
4,000,000	Early man	3,000,000*
40,000	Homo sapiens	10,000
10,000	Civilization	6,000

* As little as 15 years ago, man's origins were traced back no more than 500,000 years. But recent work being done in Africa seems to be pushing them back further and further at the moment.[1]

How carbohydrates, fats, nucleic acids, and proteins formed, is a much easier question to answer than *why* they formed. Some scientists have said life arose by chance or accident, according to the laws of probability. Isaac Asimov expressed such a view several years ago when he said,

> Once upon a time, very long ago, perhaps 2.5 billion years ago, under a deadly sun, in an ammoniated ocean topped by a poisonous atmosphere, in the midst of a soup of organic molecules, a nucleic acid molecule [DNA] came accidentally into being that could somehow bring about the existence of another like itself—And from that all else would follow.[2]

The chance or accident theory implies that the first living organism arose from non-living matter, without a pre-existing organism to make it. In other words, the egg preceded the chicken! Some scientists are now amending the accident theory, and side-stepping the chicken-egg argument, by proposing a possible "self-assembling" capacity in matter and a "built-in" tendency to form life. They point to the "unusual" properties of carbon and the other constituent elements from which life is formed, in this respect, and point out that those properties are "just right" for the formation of life. A few scientists are going even further and are claiming that life forms "inevitably" when the proper chemicals are combined, and that to some degree it is actually difficult or even impossible to prevent that formation. The inevitability, they say, is as natural a feature of the behavior of matter as perhaps the inevitable neutralization of an acid by an alkali, although tremendously more complicated.

Overall, the historical pendulum does seem to be swinging away from the accident theory toward some form of an inevitability theory. The shift can be detected, for example in the words of Dr. Cyril Ponnamperuma, a prominent origins of life researcher who in 1972 said, "People used to think that the primeval elements had to sit around in the ocean for millions of years before something happened. We now know that once the right molecules accumulated at the right time and in the right arrangement, life could begin almost instantaneously. Evolution is what takes time."[3]

Still other scientists believe life on earth may have originated from an extraterrestrial source. More and more evidence is accumulating all the time in support of this contention. For example, astronomers have identified the presence of many chemicals in outer space, among them ammonia, alcohol, and formaldehyde—the precursors of all living things.[4] Interestingly enough, several amino acids have also been detected in moon-rock samples.[5] It's becoming increasingly apparent that the precursors of life can actually be found distributed throughout the universe. As Dr. Sydney Fox of the University of Miami said several years ago, "The whole picture is one of cosmic unity."[6]

Such chemicals could have reached the earth in meteorites. Over 20 meteorites have been found in recent years containing various kinds of simple organic molecules. The most important find to date has probably been the Murchison meteorite, which fell to earth in southeast Australia on September 28, 1969. Fragments of the meteorite were picked up within hours of its fall and rushed to the NASA research center near San Francisco. Analyses revealed the presence of 16 amino acids, five of them among those found in proteins on the earth.[7] Various hydrocarbons were also found. The Murchison meteorite is rather compelling evidence that extraterrestial life is not only possible but probable. It does, however, prompt the rather strange thought that such life may be comprised of a different amino acid alphabet! The meteorite also strongly suggests that some, many, or perhaps even all of the simple chemicals involved in the formation of life on earth may have reached us in the form of "seed" material from outer space! Putting the question aside, however, of where the materials to begin life on the earth may have come from, and why they may have gotten together, let's return to the question of *how* life chemically arose on the earth.

If one had proposed a theory as to how a machine had been constructed, the best way to verify the theory would be to attempt to build the machine according to the theory and see if it worked. This is exactly what researchers are now doing with their theories about the origin of life on earth, and their ultimate goal is thus to "create life."

Serious physical research into the primordial origin of life began in 1953 when S. L. Miller made a startling discovery.[8] He passed an electric spark through a mixture of methane, ammonia, hydrogen, and water vapor continuously for one week to simulate early conditions on the earth. Several amino acids were found in the products, including glycine and alanine, the simplest amino acids, and the most prevalent in living organisms. The amounts of the amino acids formed was "surprisingly high." At a single stroke, this rather amazing experiment restored respectability to the spontaneous generation theory! Similar experiments since have produced many of the precursors of life thought to have been necessary for its formation.[9]

In the last several years, Dr. Fox and other researchers at the University of Miami have taken mixtures of amino acids and heated them under dry or semi-dry conditions. The amino acids have hooked together into long molecular chains that Dr. Fox calls protenoids.[10] They resemble contemporary protein. Placed in hot water, the protenoids quickly form clumps of molecules that Fox calls microspheres. The microspheres are more complex than the coacervates mentioned earlier. In many ways they actually resemble living cells and are even able to reproduce by division. The protenoids and microspheres are far removed from the complexity of today's living cells. But today's organisms are also far removed from their primordial ancestors! Today's proteins are too complex and too evolved to act as starting material for the origin of life. In a sense it would be like trying to produce a Model T from parts obtained from a 1978 Cadillac! The original protein and original cell must have been very simple, but with the potential to evolve. Dr. Fox believes his work is supportive of the "inevitability" theory of the formation of life. The research is continuing.

In February of 1977, a team of Japanese scientists headed by Dr. Fujio Frami, chief of Life Science Research Institute of Mitsurishi Chemical Industries, Ltd., publicly announced the successful production of a new type of "artificial cell." The cells, cultivated in sea water heated to 302 degrees F. for one month, were without nuclei or membranes, and resembled bacteria. Dr. Frami believes they are of a type which may have preceded living cells in the evolutionary process.

Meanwhile, other researchers are creating the components of living cells (enzymes, genes, proteins, etc.) by a "short-cut" procedure involving assembling "chunks" of the material involved.[11] On December 15, 1967 many newspapers carried the headline, "Scientists Create Life in Test Tube." The news stories described the fact that Dr. Arthur Kornberg and his associates at Stanford University had been successful in synthesizing the essential replicating mechanism of a small bacterial virus. The synthetic virus could utilize cell material and could reproduce itself within that cellular matrix. President Johnson was quoted in newspaper accounts as saying that the accomplishment was one of "the most important news stories you ever read, or your daddy ever read, or your granddaddy ever read."

It's debatable whether such headlines and newspaper accounts were sensationalized. A virus depends upon a living cell for its existence. The living cell carries out its replication. Whether it is "alive" and anything more than a self-replicating chemical is, therefore, debatable. Even Dr. Kornberg would not say that his virus was "alive." The research does indicate, however, that another major step has been taken toward the creation of life. Certainly few scientists would now argue with Dr. J. D. Bernal, a pioneer in launching the use of physical methods to study living material, when he said several years ago, "Life is beginning to cease to be a mystery, and is becoming practically a cryptogram, a puzzle, a code that can be broken, a working model that can sooner or later be made."[12]

In late 1973, at the annual meeting of the American Association for the Advancement of Science, Dr. Ponnomperuma echoed those words when he said, "There is no reason to doubt that we shall rediscover, one by one, the physical and chemical conditions which once determined and directed the course of chemical evolution." How close are we to creating cellular life? Probably not too close at the current level of effort. The task of producing a one-celled "living" organism will be extremely formidable, even by the short-cut method of assembling preprepared proteins, nucleic acids, etc. The difference in complexity between a virus and a one-celled organism is probably as great or greater than that between methane or ammonia and a virus. The creation of "intelligent" life would be a much greater step yet, although it should be pointed out that a one-celled organism should be amenable to the reproductive, genetic engineering, cloning, and hybrid animal techniques described earlier in the book.

To bring the whole picture into focus, it should be pointed out that in 1965 Dr. Charles Price, the newly-elected President of the American Chemical Society, publicly proposed that the synthesis of life be made an American national goal. He proposed a 20-year effort involving up to one-half of the nation's manpower. His plan involved first the synthesis of important precursors of life, then viruses, then sub-units of which cells are made, whole cells, and finally multicellular organisms. He believed it would take 20 years to reach the latter point, with another "century or so" necessary for the "creation of intelligence." His primary purpose in making the proposal was to give chemists a goal to work for, just as engineers had the moon. He also believed such research could result in synthetic organs for transplantation. But over and above all this he foresaw the creation of "new forms of life."

CONCLUDING COMMENTS

> At the present rate of progress, it is impossible to imagine any technical feat that cannot be achieved, *if* it can be achieved at all, within the next five hundred years. . . . With monotonous regularity, apparently competent men have laid down the law about what is technically possible or impossible —and have been proved utterly wrong, sometimes while the ink was scarcely dry from their pens.
>
> —Arthur C. Clarke, *Profiles of the Future*

A technique known as extrapolation can frequently be used with known data to predict unknown and even future facts and trends. The technique involves extending (extrapolating) a straight-line or curved graph, drawn from known points into an unknown area. The following chronological schedule, based on the accelerating pace of biological development described in Part I of this book, is now offered as a *possible,* arbitrary extrapolation into our biological future. It should be emphasized that the dates shown are intended to represent only possible technical achievement, not necessarily general availability of the achievements.

HUMAN REPRODUCTION

BY 1985

- Extensive development of chemical birth, fertility, and desire control agents.
- Extensive development of chemical abortifacients.
- Sperm banks with the capability of storing human sperm more than ten years.
- Artificial insemination with options of choice of sex and to some degree choice of qualities.

Artificial inovulation with both natural and foster mothers.

By 2000

An artificial womb.

The first cloned mammals.

AFTER 2000

"Baby factories."

Cloned people.

PHYSICAL MODIFICATION

BY 1985

Extensive transplantation of organs, limbs, and glands after the rejection phenomenon is overcome.

Extensive use of artificial organs and limbs.

BY 2000

Organ and tissue regeneration.

Frozen organ banks.

An extensive, although not complete, understanding of human genetics and heredity.

Limited negative eugenics.

AFTER 2000

Total mastery of human genetics and heredity. Extensive positive and negative eugenics.

Synthetic plants and animals.

"Humanants," "Humanals," and "Plantimals."

MENTAL MODIFICATION

By 1985

An expanded list of mind-modifying drugs.

Pain relief through electrical stimulation of the brain.

BY 2000

Disembodied animal and human brains (cyborgs).

Memory injection, editing, and erasure with drugs.

Extensive intelligence and personality reconstruction with drugs.

Mini-computer "boosters" implanted in the brain.

AFTER 2000

Head and possibly brain transplants.

Brain-machine, brain-computer, brain-brain chimeras.

Genetic engineering of mental characteristics.

THE PROLONGMENT OF LIFE

BY 1985

Increased power to postpone clinical death with machines and drugs.

The conquering of cancer.

The initial use of hibernation and body-temperature-lowering drugs.

The expansion of freezing research.

BY 2000

Most major diseases conquered.

Frozen organ banks.

Discovery of the cause of aging.

The power to increase life-spans 10 years or more by various techniques.

AFTER 2000

Indefinite postponement of death, and even immortality, through the development of anti-aging, freezing, hibernation, suspended animation, and sleep techniques.

THE CREATION OF LIFE

BY 1985

The production of artificial viruses.

BY 2000

The synthesis of unicellular organisms.

The discovery of extraterrestrial life.

AFTER 2000

The synthesis of multicellular organisms.

History could well prove some, many, most, or even all of the previous chronology to be poor prophesy. However, if only a small fraction of the developments listed come to fruition as predicted, the biological revolution will have to be considered at least a candidate for being the greatest scientific revolution of all time. If half or more of them become reality, there would be no contest!

Through the "pill," a mere forerunner of other developments yet to come, the biological revolution has already greatly altered history and created some staggering societal changes. This situation can only increase, as some, many, most, or even all of the events just prophesied unfold. Part II will examine the implications and possible consequences that may be on the way, and their possible role in history. The rather startling proposition will be advanced that man may be well on the way toward building another Babel—a Bio-Babel!

PART II

THE IMPLICATIONS AND POSSIBLE CONSEQUENCES OF THE BIOLOGICAL REVOLUTION

The time has come when the public must be made aware of the great impact of biological thought and knowledge. Such awareness will lead to the further liberation of man and the flourishing of his great potential. But he will be called upon to avoid the new dangers that liberation brings. Here only his sense of values can be his guide. Meanwhile the managers of society should be given some advance notice of what may be in store.

—Dr. Jonas Salk

INTRODUCTION

> Now the whole earth had one language and few words. And as men migrated from the east, they found a plain in the land of Shinar and settled there. . . . Then they said, "Come, let us build ourselves a city, and a tower with its top in the heavens, and let us make a name for ourselves, lest we be scattered abroad upon the face of the whole earth." And the LORD came down to see the city and the tower, which the sons of men had built. And the LORD said, "Behold, they are one people, and they have all one language; and this is only the beginning of what they will do; and nothing that they propose to do will now be impossible for them. Come, let us go down, and there confuse their language, that they may not understand one another's speech." So the LORD scattered them abroad from there over the face of all the earth, and they left off building the city. Therefore its name was called Babel, because there the LORD confused the language of all the earth. . . .
>
> —Genesis 11:1–9

It probably matters little whether one takes the above account literally or not, for even if there wasn't an actual, historical Tower of Babel about 4,200 years ago, the words describing it are at least 2,500 years old in themselves! And thus the more important question about the Tower of Babel account is probably whether or not it conveys a valuable, age-old, historical lesson or even warning.

The account seems to say that mankind at some time in the past initiated a massive edifice, contrary to God's will, which was to be a symbol of man's rejection of God—a monument to his own glory. However, man began to build that edifice without a clear knowledge of its purpose and possible consequences. He began to build impressively, but too soon and too fast, historically speaking, before he was really ready. God saw that the edifice was just a beginning, and that "nothing" would thereafter be "impossible" for man, and restrained from his imagination. And so God intervened, confused man's language, and scattered him, before man could bring serious conequences upon himself with his inventive genius—possibly even his own extinction.

If there is an important historical lesson to be found in the Tower of Babel account, it should be obvious that we haven't learned it yet! Even now, 4,200 or more years later, we seem to once again be at a point in history where we may have become "too big for our britches." We are once again priding ourselves as being the most intelligent, knowledgeable, practical, cosmopolitan, "advanced" men in history. And once again we are at work building massive monuments to our own glory. The greatest edifice of them all could well turn out to be the biological revolution. However, are we once again building with no clear purpose in mind and no clear knowledge of the possible consequences? Do we really yet know what we are, who we are, why we are here, and what is the purpose of human existence? Are we again building impressively, but too fast and too soon, before we're really ready? Or, is the biological revolution simply "an idea whose time has come"? Will we succeed this time in making "a name for ourselves" by actually building a tower with "its top in the heavens"? Will the biological revolution usher in man's ultimate destiny—to be as a god? Or, will God or some other historical force intervene again, and again thwart our efforts? Will we end up with simply another Tower of Babel—this time a Bio-Babel? In Part II we will examine these and other questions more closely.

CHAPTER 1

BABEL?

> It was the best of times, it was the worst of times,
> it was the age of wisdom, it was an age of foolishness,
> it was the epoch of belief, it was the epoch of incredulity,
> it was the season of Light, it was the season of Darkness,
> it was the spring of hope, it was the winter of despair.
> —Charles Dickens, *A Tale of Two Cities*

Mankind could very well be on the way to making a serious, perhaps even tragic, historical mistake. We have overlooked the obvious! We have largely failed to recognize, and thus have essentially failed to deal with, the fact that we now live in the most abnormal, most unpredictable, and most dangerous period of time in the history of mankind. And yet, despite that fact, we find ourselves in the midst of and proceeding with perhaps the greatest scientific revolution of all time: the biological revolution.

It is extremely important that the reader "feel" as well as recognize the abnormal, unpredictable, and dangerous nature of our age, for that discernment will now be the basis or foundation upon which will rest much or even most of what will be said about the biological revolution in Part II. Thus we will attempt to build that basis or foundation as firmly as possible in this chapter before returning to a specific discussion of the biological revolution in subsequent chapters.

THE JOURNEY

Someone once said that you can't possibly know where you are going unless you know where you are and where you have been. That unquestionably also applies to the biological revolution. To begin with, we'll take a historical "journey" and see how our age became so abnormal.

Our journey begins with the Greeks, for most would agree today that they were probably the first scientists in history. They were certainly the first people in history to believe that nature was understandable and explainable by man, without resorting to gods or the supernatural as part of their explanation, as did earlier peoples. The Greeks asked some very penetrating questions about nature; however, their goal in asking those questions was usually to explain *why* things happen, rather than *how.* They were thus more interested in theoretical, philosophical, religious answers than in strictly practical answers. They usually stressed reasoning, logic, and thought, rather than experimentation, as being the key to problem-solving.

With regard to the nature of matter, Democritus and others toyed with the idea that it consisted of atoms. Most, however, finally decided that all matter was simply comprised of some combination of four elements (fire, water, earth, and air) and four qualities (hot, cold, wet, and dry), as Aristotle and others had contended. Aristotle and his followers had claimed that one form of matter could be converted into another if its combination of elements and qualities could somehow be changed. This contention, coupled with man's timeless greed, led to the establishment of alchemy about the time of Christ. Alchemy flourished in the Near East for about 1,200 years and then in Europe for over 400 more. The primary goal of all alchemists was to somehow change the base metals into gold and silver.

Both alchemy and early science sprouted from the intellectual seeds which the Greeks had sown. But, fed by greed, it was alchemy which grew, crowding out and stunting the growth of early science. The Romans might have weeded alchemy out of man's intellectual garden if they had been more interested in scientific matters, but they were not. They were statesmen, engineers, soldiers, but not scientists. They were content to refer to the writings of the Greeks for any scientific information they desired. And so early science withered, for lack of nourishment, while alchemy flourished.

The Greeks had strongly pushed the pendulum away from experimentation toward reason, logic, and thought in problem-solving. The pendulum remained essentially stuck there for almost 2,000 years. However, the alchemists did have to experiment in attempts to achieve their theoretical goals, and the pendulum began to vibrate. About 1200 A.D., revolutionaries like Roger Bacon began to appear who advocated experimentation as being the key to problem-solving. Bacon was followed by Copernicus, Keppler, Galileo, and Boyle, men who all produced increasingly significant historical pushes on the pendulum. The Renaissance, the end of feudalism, and the discovery of the printing press also provided indirect but significant pushes. To make a long story short, the pendulum finally swung to a new way of problem-solving in the 1600's. In the new method, the techniques used were usually just the reverse of those used by the Greeks. Generalizations were considered to be only imperfect representations of the real world and were thus derived from observation and experimentation rather than simply logic and thought. Experimentation was stressed as being the key to problem-solving, rather than thought, and more emphasis was placed on finding out *how* than *why.* This new method, which has been used ever since, became known as the scientific method. It has produced over 300 years of scientific and technological success stories, culminating thus far with men walking on the moon on 1969 and machines landing on Mars in 1976.

For about the first two-thirds of those 300 years, scientific and technological progress came rather slowly, in a rather unpredictable and uncoordinated way. The scientific method was not a dramatic, overnight success! It was almost as if man had set out on a long, new journey. The scientific method was his vehicle, but it was a vehicle without a motor as yet. It had to run on muscle, water, and wind power, for it was still a horse and buggy age.

But in the early nineteenth century, a new epoch began. Man's capacity to do ceased to be limited by his own strength and that of the men and animals he could control. He traded his vehicle in for a scientific method powered first by the steam engine and the Industrial Revolution, and then later in the century by electricity. With the installation of a motor, the vehicle began to accelerate. For example, in the late 1890's, Charles Steinmetz pointed out that the first 60,000 kilowatt generator could produce more energy than the total slave population of 1860 combined!

In the twentieth century, powered additionally first by the internal combustion engine, and then later by nuclear power, the scientific method really began to "roll." Man began to solve greater and greater problems, even global problems of food production, disease, and transportation, and he has moved faster and faster down what has appeared to be a straight, smooth highway of science and technology.

An understanding of the historical material just presented lays a foundation upon which to rest five extremely important historical observations about our modern scientific age.

1. Historically speaking, the modern scientific age began with a rather sudden "swing of the pendulum"!
2. Historically speaking, the modern scientific age has been with us for a relatively short time—for only about 300 years. Interestingly enough, if the three million year history of man were written in a three thousand page book (one thousand years of history per page) the modern scientific age would have to be described in the last paragraph of the book!

3. Of the 300 years or so of "progress" which the scientific method has produced, most of it has actually occurred in the last 75 to 100 years. This period of time would have to be described in the last sentence of the afore-mentioned book! Or, if human history were to be represented by a 24-hour clock, most of our scientific and technological progress has occurred in the last couple of seconds before midnight!
4. For thousands of years, religion was the predominant influence in man's life, but it has been replaced in the last 75 to 100 years by science, technology, and the machine. We now live in an age where our lives are predominantly controlled by the increasing, accelerating change produced by science and technology.
5. In terms of scientific and technological growth, the last century or so has been by far the most abnormal period in the history of the earth. However, it has appeared to be normal to us because we have lived entirely within it and we thus have no basis for comparison. Scientific and technological growth in the United States has been particularly abnormal. In 1850 for example, after a visit to the United States, the famous French writer de Tocqueville commented, "Among the civilized peoples of our age, there are few in which the highest sciences have made so little progress as in the United States."

These five observations should have been obvious in the twentieth century, but for the most part, they have not. Perhaps we have become so intoxicated with our acceleration and speed that we have forgotten to some degree where and how our journey began, and how far, historically speaking, we have come.

SCIENTISM

There are two extremely important, interrelated questions which the previous five observations almost automatically prompt. First of all, are we as well adjusted to our scientific and technological age as most seem to think? And secondly, is it possible that we may actually be far more maladjusted than we realize? For various reasons (including ignorance of the five previous observations), society has not asked these two questions as diligently as it should have. Even so, the questions have prompted the development of a whole spectrum of thought. The spectrum can be most clearly defined by examining its extremes or ends.

One extreme to the spectrum is as old as the scientific method itself. In fact, ever since the time of the ancient Greeks, there have been those who have believed that the search for knowledge and truth should remain man's noblest goal, and it in itself *completely* justifies the practice of science. Those who see science and the scientist in this light have always believed that if man is to progress, the search must go on. They have envisioned the highway of science and technology as remaining straight, smooth, and endless, truly leading man into a golden future and his golden age. This viewpoint can be traced back at least as far as Benjamin Franklin in the United States, In a letter to Joseph Priestly, Franklin once remarked, "The rapid progress true science now makes occasions my regretting sometimes that I was born too soon. It is impossible ot imagine the height to which may be carried, in a thousand years, the power of man over matter."

One has to wonder to what extent Franklin would have been impressed with what has happened just 200 years or so later. Could he, for example, have foreseen in even his wildest prophetic dreams the developments of the 1960's where knowledge doubled, the genetic code was deciphered, attempts to create life were underway, organ transplants were initiated, many new subatomic particles were discovered, computers drastically reduced the work load, and man walked on the moon and began to dream of space colonies?

Is it any wonder we have an extreme today on the spectrum in which many or even most believe that science should not be just a limited method for finding truth, but a universal ethic. The thought is deeply rooted that *only* science and technology can now solve the problems of our age and of the future, and that given enough time, money, and effort they can indeed do just that. This widespread extreme in thought has become known as *scientism* in the last generation or so. Scientism is an intoxication with change, progress, and an optimistic future, as can be seen in the following two quotations, the first by G. Eastman, the second by Emmanuel Mesthenes.

> Scientism . . . assumes "science" designates *the* true and ultimate way to solve the problems of nature and man. It is, in fact . . . a de Godized religion. . . . It is the mid-twentieth-century version of eighteenth-century deism. Like the deists, practitioners of scientism have denatured God, but deified nature, assigning to scientism an omnipotent role of solving all problems, clarifying all things.[1]

> We have now, or know how to acquire, the technical ability to do very nearly anything we want. Can we transplant human hearts, control personality, order the weather that suits us, travel to Mars or Venus? Of course we can, if not now or in 5 or 10 years, then certainly in 25 or in 50 or 100.[2]

The rallying cries of scientism are, "Full speed ahead!"; "Our superhighway is straight, smooth, and leads to the Golden Age of Man!"; and "On to the stars!" Some say that our safety is now in our speed. If we were to slow down, stop, or attempt to go back from whence we came, they say, history would pass us by, or we could be struck from behind by the problems we have already passed, or by other nations attempting to go around us.

There is no shortage of prophets in this respect either. Some believe we are moving toward a brave new world in which human beings will succeed by becoming more mechanized. Harvey Cox has predicted the coming of "megalopolis" in a "technopolitan era." Marshall McLuhan has predicted the development of a "global village." Jacques Ellul has foreseen a new era of "technological society." J. K. Galbraith has predicted a "new industrial state" dominated by the "technostructure." Daniel Bell and Amitai Etzioni have looked even further into the future. Etzioni has foreseen the coming of a "post-modern period" and Bell, in his book of the same name, has predicted the coming of a "post-industrial society" in which intellectual activity will replace primary production as the basis of economic life. Many claim Teilhard de Chardin has looked furthest of all, perhaps to man's ultimate destiny. Chardin has predicted an eventual metaphysical union of matter and spirit.

How deeply is scientism entrenched among laymen? The National Science Foundation sponsored a poll taken of 2,209 people (statistically extrapolatable to the entire United States population) in May and June of 1972. The poll sheds at least some light on our question. In response to the question of whether science and technology have changed life for the better or worse, 70% said "better," 8% said "worse," and 11% said "both." When asked what words described their general reaction to science and technology, 45% said "satisfaction or hope," 23% said "excitement or wonder," while "fear or alarm" was selected by only 6%. When asked about the net impact of science and technology on our world, 54% felt that, overall, "science and technology do more good than harm," 4% said "more harm," and 31% indicated "about the same." (Note: deviations from 100% in the questions were due to "no opinions.") From a list of nine professions, that of scientist was ranked second only to physicians in terms of respect. It would appear that scientism among the general public is fairly widespread, at least based on the results of this and similar polls.

THE ANTI-SCIENCE MOVEMENT

The results of the poll just mentioned have been used by some to stress the point that a majority of people still react favorably toward science and technology. The most significant thing about the poll, however, could well be the feelings of the minority. Upwards to 8% of those polled expressed strong negative feelings about science and technology, upwards to 31% had reservations about their value, and upwards of 22% had no opinion. These figures become impressive when one realizes that until the end of World War II they would have undoubtedly been much lower, perhaps even to the point of being negligible. There were a few dissident voices critical of science and technology before 1945, but they were usually simply drowned out by the vast majority. Why the rather significant shift in opinion?

In 1945, man hit the first bone-jarring, teeth-rattling bump or vibration in his superhighway of science and technology. What a shock it was to suddenly learn that with the atomic bomb he held the power to erase himself as a species from the face of the earth—all with the push of a button! That thought produced feelings ranging from uneasiness to outright fear in many. Since 1945, the bumps and vibrations have continued and have seemingly been coming more frequently. We have had serious problems with detergents, nuclear testing, oil spills, smog, pollution, the weather, the pesticide DDT, the herbicide 2,4,5,-T, the SST, energy, metals and minerals, and food supplies, to mention just a few. It's been one problem after another since the mid-1950's, with the products of science and technology always directly or indirectly involved. Our highway into the future has proven to be nowhere near as straight and as smooth as we once thought. This knowledge has generated an increasing chorus of dissatisfaction with science and technology, from both within and without. Aren't they the primary influences which now mold and control our lives? Who else should we blame, many ask? The dissatisfaction has increased to the point that another extreme has thus appeared in our historical spectrum of thought. Generally speaking, it is an extreme characterized by a feeling that we have been deceived and perhaps even betrayed—a feeling that, overall, science and technology have created more problems than they have solved and have done more *to* man than *for* man. Samuel Silver summarized this new extreme in thought when in 1969 he said,

> There is a feeling, which is growing in the United States and in other western countries, that the advances made through science and technology have somehow failed their purpose; that the hope placed in them by mankind for the attainment of a more satisfying life and of a happier and more tranquil world has suddenly been betrayed. There is in consequence a growing sense of dismay and frustration regarding science and technology.[3]

Other defectors from scientism now include leading writers, scientists, and philosophers. Men such as Lewis Mumford, Herbert Marcuse, Theodore Roszak, William Irwin Thompson, Rene Dubos, Barry Commoner, Bruce Catton, Rachel Carson, Ralph Nader, Archibald MacLeish, and others have to varying degrees accused science and technology of being both fundamental causes as well as curers of humanity's major problems today. Their common concern seems to be that science and technology frequently create serious problems which only science and technology can then solve, with the solutions often producing additional problems. Man, they say, is all too often now caught in a vicious cycle produced by his own ingenuity. And yet, only his ingenuity can save him!

Anti-science feelings obviously now run very deep in many people. Coupled with anti-war and anti-capitalism beliefs, they have surfaced significantly at least twice in rather militant forms. In 1969 and 1970, a series of activists, protestors, hecklers, and demonstrators repeatedly harangued and disrupted sessions of annual meetings of the American Association for the Advancement of Science (AAAS). Individuals describing themselves as affiliated with the "Science for the People" movement and the "Peoples Science Collective" claimed this was their only opportunity to tell scientists that the work they do is "being used by and large to destroy people, to corrupt the environment, and to suppress and oppress minorities." Ironically, the AAAS programs at those meetings contained more sessions dealing with contemporary social ills than ever before!

Trying to specifically characterize the new anti-science extreme in thought can be rather difficult, for it is comprised of an overlapping mixture of charges, with different charges or combinations of charges stressed to varying degrees by different persons. About all one can do, therefore, is to look at some of the individual charges that have been leveled at science and technology.

For example, there are those who look at the nuclear arms race with growing dismay and fear. They point out that man has probably been at peace less than 10% of his existence, despite thousands of treaties, and that we do not understand whether our tendency toward conflict and waging war is instinctual, or a learned behavioral response, or both. Therefore, they ask, how long can it be when children play with loaded guns before those guns go off? Weapons are obviously not evil in themselves, nor are their manufacturers necessarily evil. But those who buy them and use them frequently are, as evidenced by history. Therefore, are the developers, manufacturers, and suppliers of weaponry at least somewhat responsible for how they are used, knowing that sooner or later they will probably inevitably be used in an evil way? It almost seems at times that we will have to either eliminate science or eliminate war, for we apparently can't have both!

Critics of our automated machine age use similar arguments. Machines are obviously not dehumanizing in themselves, but when employed by men they can become that. The late Walter Reuther once said, "The prospect of tightening up bolts every two minutes for eight hours for thirty years doesn't lift the human spirit."

He overstated his case, but there is a ring of truth in the claim that man has become subordinated to the machine, rather than the other way around. And if we have substantially put our faith in the machine rather than in ourselves, Reuther's statement may be more realistic than we would like to admit.

Is the machine age stifling man's instinctual needs of expressing his strength, reflexes, creativity, and competitiveness? Is it stifling his inner needs of recognition, a sense of achievement, a sense of responsibility, and growth? Lewis Mumford, Harvey Cox, Bruce Catton, Rene Dubos, Jacques Ellul, and others have answered these questions in the affirmative. They believe machines are dehumanizing us. For example, Mumford, in *The Myth of the Machine,* recommends that we overthrow the "megamachine" before it overcomes us.

He claims that the consequences of mass production, automation, and computers outweigh their benefits. Harvey Cox, in *The Feast of Fools,* states his belief that "technology has damaged the inner experience of man." Bruce Catton, in *Waiting for the Morning Train,* suggests that it may be time to ask an even more far-reaching question. Does man still control his own destiny or is that merely an illusion? He suggests that the machine age "operates only at full speed," but, "unfortunately, it cannot be steered."

Is the materialism that science and technology have produced resulting in increasingly hollow, empty, meaningless, impersonal, purposeless lives today? Are *things* now in the saddle, riding us? Certainly people are healthier and have more things than ever before, but have these facts made us better men? Lewis Mumford has asked these questions even more pointedly.

> Why has our inner life become so impoverished and empty, and why has our outer life become so exorbitant, and its subjective satisfactions even more empty? Why have we become technological gods and moral devils, scientific supermen and esthetic idiots—idiots, that is, in the Greek sense of being wholly private persons incapable of communicating with each other and understanding each other?[4]

Is Mumford right? If so, *why* have we become "scientific supermen and esthetic idiots"? Could at least part of the answer lie in the fact that modern science is now geared to determining *how,* and usually ignores *why?* But is *why* also important for intellectual satisfaction? Does an emphasis only on *how* lead to a cold, intellectually incomplete materialism? Rene Dubos, Eugene Wigner, and W. T. Stace are among those who have stated their belief that such is the case.

> The most obvious reason for disenchantment is the realization that prosperity and comfort do not assure health and happiness. In fact, material progress often has consequences that spoil the quality of life. . . . Science, it has been said, gives man everything to live with, but nothing to live for.[5] —Rene Dubos

> We owe many of our comforts and much of our leisure time to progress, but it has made life so easy there has been nothing to strive for. It's wonderful to have things soft—but if everything is soft, life has no purpose.[6] —Eugene Wigner

> If the scheme of things is purposeless and meaningless, then the life of man is purposeless and meaningless too. Everything is futile, all effort is in the end worthless. A man, may, of course, still pursue disconnected ends, money, fame, art, science, and may gain pleasure from them, but his life is hollow at the center. Thus the dissatisfied, dissillusioned restlessness of modern man.[7] —W. T. Stace

Some believe that the hollowness of our age may also be at least partly caused by our overexpectations. After about 300 years of scientific success stories, we almost automatically began to place a high level of faith in science and the scientific method—even to the point of considering them to be omnipotent and even infallible. But that faith has now been shaken to some degree as we have begun to realize that science and the scientific method cannot solve all of our problems. There are limits to their effectiveness, as Phillip Abelson has pointed out: "The public needs to understand that science and technology cannot be applied successfully to the fulfillment of every wish."[8] But as we have realized that our expectations go beyond what science and technology can produce, have we entered an age characterized by increasing fear, disappointment, frustration, and depression, with those traits moving lower and lower in the age structure of society? Is that a major reason for the burgeoning interest in such non-scientific areas as astrology, mysticism, and witchcraft today? Are those who used to place unrealistic expectations in the scientific method now taking those expectations elsewhere?

Many critics of our age believe we are also being overwhelmed by mounting confusion caused by an inability to cope properly with a growing mass of knowledge. Has the twentieth century become an age of confusion? Have we entered an age where an accelerating production of information and its growing mass now preclude anyone being an expert anymore? Are we being buried under a rising mountain of knowledge?

Knowledge is said to have doubled from the time of Christ to 1750, doubled again from 1750 to 1900, doubled again from 1900 to 1950, doubled again in the 1950's, and doubled again in the 1960's. Some believe that at the present time it may be doubling every seven to ten years. At present, there are over two million items of information (books, journals, articles, etc.) appearing worldwide per year. There are about 400,000 books and over 100,000 technical journals published worldwide alone! Many predict there could be 12 to 14 million items appearing per year by the year 2000 if research funding and the number of scientists available continue to increase and the literacy output per scientist continues to increase also.

Consider the plight in just the biological and medical sciences at the present time. There are about 6,000 biological and medical journals now being published per month. Assuming 100 pages each and 500 words per page, that's 300 million words per month that someone in those fields would have to read and digest each month to just stay current with everything being printed in his field. Reading at a rate of 200 words per minute, continuously for 24 hours per day, he could read only about 3% of it in a month's time! Is it any wonder that scientists today specialize in some sub-field of a sub-field within their field? And most find they can't keep up with even that!

Is a flood of uncoordinated change, brought on by a roaring, rising torrent of knowledge and information, producing a restless, scientific society in which we all feel uprooted and bewildered to at least some degree? How great a rate of knowledge bombardment and resultant accelerating change can man tolerate? It has been said that we only use about 10 percent of our mental capacity. But may it not require a gradual, long-range evolutionary adjustment to increase that figure? What happens in everyday life when we overload that 10 percent? On a daily basis, might the 10 percent figure simply be a confusion threshold? Archibald MacLeish has called for a moratorium on new science and new research until we can cope with the enormity of the information we now have. Does he perhaps have a point worthy of consideration?

We seem to be caught in a serious knowledge gap today. Most scientists can't even keep current with what is happening in their own isolated corner of their own isolated field, let alone keep up with the rest of the sciences and the other disciplines. But, most important of all, we have also largely lost the ability to see the global "big picture," which involves all areas of knowledge. This is extremely unfortunate because our problems and concerns are no longer the problems and concerns of simply individual nations and geographical areas but are now concerns and problems of the whole human race. The prospect of *national* and *regional* disorder and chaos has now been joined by the prospect of *global* disorder and chaos. As U.N. Secretary-General Kurt Waldheim said in his State of the World speech on September 5, 1974, "The choice is between order and chaos. From time to time, throughout history, these alternatives have confronted societies at times of change. The only difference today is that this choice does not apply to one nation or civilization but the whole world."

Mankind used to be protected by a national or regional awareness that there were some things we couldn't or shouldn't do. But now that science and technology are operating on a global basis, that awareness has been blurred or has even disappeared because of our inability to see the "big picture." Science and technology have turned the world into a giant spider web wherein a disturbance in one area usually now disturbs the rest of the web as well. It has become difficult to predict the outcome of our experiments now that we have turned the whole world into our laboratory and the human race into our subject! We seem to have entered an age of surprises where the final outcome of our experiments is frequently much different than what we expected.

It almost seems as if there has been one basically unforeseen, surprising global problem or crisis after another since 1955 or so. For example, there has been the fallout problem, the nuclear arms race, the deterioration of the biosphere, the DDT problem, possible inadvertant weather modification, the thalidomide generation, the rise of antibiotic-resistant bacteria, and technological failures leading to psychological stress in larger cities, to mention just a few.

In the 1970's, we have added the energy, food, and weather crises to our list. We have also been shocked to find that many of our everyday chemicals (e.g. the plastic, polyvinyl chloride), which we formerly considered to be relatively safe, must now be placed under suspicion of being extremely dangerous in even trace amounts. We are even beginning to suspect that our radio and television signals and the electro-magnetic fields generated by the transmission and use of electricity may be more harmful than we had thought. But the most shocking surprise of all involves the recent suspicion that the fluorocarbons used in countless spray cans around the world may be partially destroying the ozone layer high in the atmosphere that protects us from the ultraviolet light reaching us from the sun. If it turns out that even our deodorants, hair sprays, and shaving creams can alter the viability of the earth and endanger man, it will indeed have proven to be a surprising age!

We are slowly coming to the chilling realization that, because our problems today are so complex and so widespread, it is possible that we could seal our fate before any dramatic symptoms appear. Kenneth E. F. Watt described this extremely serious possibility recently when he said, "A still more fundamental problem is that whereas formerly a civilization that erred would not have been powerful enough to prevent the rise of later civilizations, this is no longer true."[9]

It should be obvious that something has gone wrong in society. But what? To what extent are the nuclear arms race, the machine age, materialism, overexpectations, confusion, the knowledge gap, and an age of surprise crises contributing factors in our societal problems? Have science and technology reached the point where they are doing more to man than for man? Has man created his own inevitable fall by eating from the tree of knowledge? Have we overeaten? Is our appetite out of control? Or, are science and technology now our only hopes for survival in the future?

We live in an extremely paradoxical age. On the one hand, we are asking some extremely serious questions about the value and the dangers of science and technology and are beginning to question the idea that only more science and technology can solve our problems. And yet, on the other hand, we continue to maintain that they may be our only hope for the future. Some continue to talk about moving into the Golden Age of Man while others predict possible global catastrophy and collapse within the next 30 to 100 years! The scientism and anti-science ends of the spectrum are both so extreme that they both appear to be equally naïve and unrealistic. Thus most people are simply confused. They don't seem to know where they are on the spectrum —or where they should be! We can no longer rely only on science and technology to solve our problems, but neither can we reject all things scientific and technological. It truly is a paradoxical age! Charles Dickens' words were indeed probably never more appropriate than they are today: "It was the best of times, it was the worst of times, . . . It was the season of Light, it was the season of Darkness."

THE CLOUDY CRYSTAL BALL

One need only listen to the radio, watch television, or read the headlines today to realize that our societal destination is no longer as clear as we once thought. But what are we to do about it? It appears that we may have reached a critical speed or degree of momentum down our superhighway of science and technology wherein it could be quite dangerous to either accelerate further or throw on the brakes. In either case we could lose control and break our necks! There may thus be only one possible alternative open to us—to stabilize our speed, keep an eye on the speedometer, move ahead cautiously, and erect signs for those yet to come, saying, *Caution—Science and Technology Ahead!* Since the early 1960's, an increasing number of people have been advocating just such a middle-of-the-road (or middle-of-the-spectrum) policy. The watchword or rallying cry for this growing segment of society soon became "technology assessment."

In 1967, Congressman Emilio Daddario and others introduced a bill in Congress to establish a governmental Office of Technology Assessment (OTA). The bill's proponents fought an uphill fight of almost five years, against strong opposition from many quarters, before finally getting it passed in 1972. Unfortunately, however, OTA's accomplishments and progress since have been less than spectacular.

One of the bigger reasons for OTA's inertia to date involves the fact that technology assessment is currently much more of an art than a science—and probably always will be. There are currently at least six major methods available for its practice, with each method having its own group of practitioners and proponents. Four of the methods (extrapolation, intuitive forecasting, scenario writing, and Delphi Forecasting) are known as "trend" methods. The other two (morphological analysis and decision or relevance trees) are known as "normative" methods. It really isn't necessary to define any of these methods in depth, for it can simply be stated that there is no set, single way of conducting technology assessment, nor any combination either, that has proven satisfactory. It truly is only an art rather than a science—and a rather confusing art at best. As Raymond Bauer has said, "How does one carry out technology assessment? I suppose that at this stage the problem is akin to that of how one can eat an elephant. The only answer is that one must begin by biting the elephant. And, considering the magnitude of the task, it is difficult to argue that one place is better than another for biting to start. And, after a considerable amount of biting has taken place, the elephant remains largely unscathed."[10]

There are also other major obstacles standing in the way of effective technology assessment, such as high costs, long periods of time needed, problems with assessor objectivity, conflicts in human values, the possibility of assessment turning into arrestment, the possible need to assess the assessors, etc. But the biggest roadblock of all—the basic unpredictability of the future—is unfortunately also the roadblock that will be the most difficult (if not impossible) to overcome. It should be obvious that a cloudy crystal ball will remain a cloudy crystal ball no matter how much it's polished! What major forecast of the future, made in the past, has ever scored a bulls-eye? Have not such forecasts always fallen far above, below, or wide of the mark? Are we doing any better today? Take television, for example. In 1948 there were 10,000 television sets in the United States. Who would have predicted the over 60,000,000 sets that are in use today? The consequences of so many television sets have been so complex, interwoven, and far-reaching that even today the sociological and psychological questions involved are still easier to raise than to answer. Or what of the Aswan Dam? It fulfilled its purpose of doubling the availability of electricity in Egypt, but also brought a series of completely unforeseen but critical geological, health, and cultural problems with it. The SST could also be mentioned in this respect. The battle over the SST was waged on several grounds, but energy consumption was not one of them. This aspect of the argument was essentially overlooked! In fact, even today only a few people realize that several hundred SST's could have consumed the jet fuel fraction of all U. S. petroleum reserves in under a decade. More recently, the catalytic muffler was designed to decrease the levels of certain atmospheric pollutants. But, after being placed in operation on millions of automobiles, it was found to actually increase the emission of dangerous sulfur compounds. Major surprises in our age are no longer confined only to technological developments. Very few prominent, long-range economic planners foresaw the energy crisis, the oil embargo, and the quadrupling of oil prices—unforeseen developments which have literally shaken the world to its very foundations.

Unfortunately, scientific and technological developments today may also contain delayed as well as immediate surprises—long-range surprises that may not show up until a certain long-range period is up. For example, when Paul Mueller was given a Nobel prize in 1945 for his discovery that DDT could be used in nature, DDT was called "the ideal pesticide." It has since been banned in the United States! Mark Twain's selective words of long ago may now generally apply to most developments today: "Soap and education are not as sudden as a massacre, but they are more deadly in the long run."

And, of course, what about the sudden, unforeseen discoveries or discontinuities which always seem to have a way of historically rearing their completely unexpected heads? As mentioned earlier, x-rays, nuclear energy, radio, television, quantum mechanics, relativity, electronics, penicillin, the Van Allen radiation belts, atomic clocks, and transistors were all basically unexpected, unforeseen discoveries. To what extent has just the unforeseen transistor changed the world since its discovery?

Although the use of crystal balls, the study of sheep livers and the movements of the stars, and the pronouncements of oracles have all been replaced by the computer and other sophistications, the future probably remains as difficult to assess as ever, for it too has become far more sophisticated. Technology assessment will probably always face the dilemma which Herman Kahn, Director of the Hudson Institute, succinctly summarized recently when he said, "The basic problem in its simplest terms is that the future is unknowable. . . . The uncertainties dominate."[11]

Even if all of the problems we have now discussed are overcome and an assessment is successfully carried out, two more problems can then result. First of all, what will happen to that assessment? It should be obvious that technology assessment can only be an advisory tool, for if it were regulatory, the assessors and the assessment could both quite easily lose both their neutrality and their credibility. Technology assessment must thus be a tool for decision makers, but not a decision maker in itself. On the other hand, however, if there is no political followup, then the assessment has served no purpose. If it only results in paper shuffling and the grinding out of unused reports, it will soon lose its credibility also. Secondly, even if followup does occur, of what value is it unless it is global in scope? For example, when the United States decided to forego further development of the SST, creating economic problems for Boeing, the Seattle area, and even the U. S. itself, France, Britain, and Russia proceeded with their plans to develop the aircraft.

Interestingly enough, even the term "technology assessment" itself creates one last problem which should be mentioned. It should be obvious that technology assessment actually involves economics, industry, labor, political science, and human values. The term "technology assessment" is thus actually limiting, confusing, misleading, inappropriate, and a misnomer! It is a disciplinary term representing what is almost always an interdisciplinary study. Unfortunately, the term has been used long enough that it has "stuck." The substitute terms that have been proposed to take its place (e.g. social impact analysis) have always seemed bland and unfamiliar by comparison. But even more important, the self-focusing nature of the term has withdrawn attention from another very important consideration. It is scientific discovery which makes technological development possible. Although science and technology are now virtually inseparable, however, very little attention is being turned toward the need for *science assessment.* Isn't science assessment at least as important as technology assessment?

There does seem to be at least some sentiment currently for the establishment of a "science court."[12] In such a court, according to those currently proposing it, "case managers" would be selected for each side of a certain scientific issue. They would compile non-value laden facts as their side sees the issue, and would then argue their case in an adversary hearing before a panel of scientific judges. If an agreement could not be reached through mediation, the judges would write opinions on the contested statement. The word "court" is thus a misnomer, for it would be left for other agencies to apply the facts and opinions tabulated, to policy decisions.

The OTA and the proposed science court are facing such a wide array of formidible problems that it is difficult to generate a great deal of confidence in their chances of significantly molding the future. Even those most optimistic about the eventual success of either or both concede that both face an uphill battle.

In the meantime, whether out of impatience, lack of confidence, or both, a whole series of individuals and organizations have turned would-be prophet. There is a new intellectual pursuit abroad in the land, although it has yet to gain an accepted name. Some call it futurism, others call it futurology, futuristics, futuribles, prognostics, etc. The pursuit has resulted in a host of books, including several best-sellers, several journals, and over 10 major councils, institutes, and centers for studying the future. The word "prophet" has been polished up and is perhaps less tarnished than it has been for centuries! As Robert L. Kuhn said in the February 8, 1975 issue of *Plain Truth* magazine,

> Prophecy is no longer a dirty word. . . . Ten years ago you would have been labeled a nut, a crackpot, a charlatan, a weirdo—for so foolishly or naïvely dabbling at the edges of society. Today, you could be a scientist, an economist, a psychologist, a theologian—a highly respected member of your profession, investigating the forefront of human knowledge. You are a prophet. You predict the future. . . . The last quarter of twentieth century, as we approach the year 2000, will herald the greatest avalanche of prophetic statements, utterances, and proclamations that the world has ever seen. As illogical and irrational as it would have seemed to a "logical" and "rational" society a few years ago, this is now coming to pass.

Overall, the current deluge of prophecy may be simply clouding rather than clearing the future. The deluge may be simply confusing our already confused and dangerous age. But our crystal ball has grown increasingly cloudy and we are now only dimly perceiving the future. Do we have any choice but to prophesy?

In summary, there are five points with regard to technology assessment, a science court, and futurism which should be emphasized. First of all, we have no choice but to attempt them. Secondly, the task is far more complex and difficult than it looks at first glance. A full-scale appraisal of the future must not only ask, but, also in some way answer, many extremely complex and difficult questions. And yet, thirdly, those doing the appraising can probably never be completely objective; they will be using a methodology which will probably always remain more of an art than a science; and they are trying to define a whole or big picture which is greater than the sum of its parts. Fourthly, because the future always has been and probably always will be basically unpredictable, the reliability and usefulness of any appraisal will always be open to at least some degree of doubt and thus must be accepted with a certain degree of faith. And fifthly, for the previous two reasons, *technology and science assessment and futurism can be practiced by almost anyone who has done his homework, for the future is essentially unknowable!*

FUTURE SHOCK

At the risk of being repetitious, there are three extremely important points about our age that deserve restatement.

First of all, as previously pointed out, we have reached an awesome, perhaps even excessive rate of speed down our superhighway of science, technology, and knowledge. Not only have nature and the rest of the environment become a blur, but so has everything else. Our speed has produced a restless, scientific and technological age of staggering, bewildering, mind-boggling, overwhelming, global confusion, complexity, and extremely rapid change—and we're still accelerating! As James Reston said in 1964, "In a restless, scientific society we are all bewildered. All the relationships of the nation, region to region, Federal government to state government, state to city, town to village and farm, employer to employee, white man to black man, and even parent to child—all these relationships have changed faster than human beings know how to change."[13]

The variables and problems of our age are already so numerous and so complex, so interrelated and interconnected, and the rate of change and flux so great, that no one can seem to get his bearings anymore and see the "big picture." We seem to have entered an age of intellectual entropy. There are no "experts" anymore, only those who have devoted more time, study, and effort to certain aspects of our problems than have others. However, such people now rarely agree on the exact nature and causes of those problems, let alone on the solutions. As a result it is also an age of great extremes in thought. The so-called "experts" can be 180° apart on many questions—and usually are—while agreeing on others. It has thus become quite difficult to categorize what the "optimists" and "pessimists" are saying, because it has become difficult to even define what an "optimist" and a "pessimist" are.

The current absence of "experts" is probably also largely responsible for the current absence of dynamic, forceful, and even significant leaders (and heroes) on both a national and global basis. Where are the men capable of "taking the reins" and leading man out of his problems and predicaments? Where is the modern Abraham, Joseph, Moses, Caesar, Napoleon, Washington, Lincoln, Churchill, Eisenhower, and De Gaulle? In the last decade or so in the United States, at least part of the answer can be found in the fact that many, most, or perhaps even all of those men with the greatest potential of becoming great leaders, or even national heroes, became victims of our abnormally complex age. Their developing images and careers were terminated by gunfire and/or infamy.

Secondly, not only has the landscape become blurred because of our high rate of speed, but so has our destination. Our crystal ball has grown cloudy, and the more sophisticated methods of prognostication that we have developed in recent years don't seem to work any better. So we also now seem to live in an age of no significant prophets, as well as of no significant "experts," leaders, or heroes.

In a sense, the future might now be compared to a giant, fantastically complex puzzle, comprised of a series of such interlinked, interrelated, overlapping, contributory sub-puzzles as population, resources, pollution, industrialization, and food production. In turn, each contributory sub-puzzle is comprised of such sub-sub-puzzles as politics, social traditions and trends, the gap between the rich and the poor, various other economic problems, and weather conditions. These contributory puzzles are also comprised of contributory puzzles. There are different ultimate pieces to the contributory puzzles for different nations and for different global regions, and these ultimate pieces can be inserted into the puzzle optimistically or pessimistically—in a sense, up or down! There are also ultimate pieces that can be inserted to form a short-term perspective in the "big picture," and an optional set of much more numerous, much harder to put together, ultimate pieces that can be used to form a long-term perspective involving future generations.

As just one example of the puzzle analogy, it might be pointed out that the nuclear waste problem is an extremely complex puzzle all in itself because of the many "pieces" involved. And, of course, most of those pieces can be inserted differently for different nations and geographical areas, optimistically or pessimistically, and to form a short- or long-term perspective (both of which may soon involve terrorism). Yet the nuclear waste problem is only a contributory puzzle in the nuclear energy puzzle, which is only a contributory puzzle in the energy puzzle, which is only a contributory puzzle in the resource puzzle, which is only a contributory puzzle in the global "big picture." It should be obvious why we never seem to get very far in putting the overall puzzle together anymore, and never seem to get to the point of seeing the "big picture." There are just too many pieces to the puzzle today, too little assembly time available, and the puzzle keeps changing! On the rare occasions when we do seem to get the whole puzzle together, or come close, it's usually because we haven't used the more complex, optional set of ultimate pieces that would have given us a long-term "big picture." And thus, our crystal ball has indeed grown cloudy, so cloudy that we are now moving into the future with very little sense of direction.

Thirdly, our high, perhaps even excessive rate of speed, and inability to clearly discern our destination are creating serious psychological as well as physical problems. We've begun to realize that something has gone wrong with our "journey." It's just not producing the peace, happiness, and contentment that we formerly always thought it would, and this realization seems to be making us increasingly uncomfortable.

Much of our uneasiness can probably be traced to the fact that, intoxicated with our acceleration, we have simply invariably ignored the warnings about excessive speed that began to appear in 1945 with the atomic bomb. As a result, in the last generation or so, we have been encountering an increasing number of surprising, unexpected, bone-jarring, teeth-rattling "bumps" and "vibrations" in our superhighway. Unfortunately, these "bumps" and "vibrations" have all too often turned into crises—crises which have already produced an 80th generation unlike any of the previous 79 that have occurred since the time of Christ. We are now faced with *global* crises that threaten the future of humanity, as opposed to regional and localized crises which the globe could afford in a sense in the past. Civilizations which erred in the past could not prevent the rise of future generations, but today they can. And apparently we may now at least partially seal our global fate before any symptoms appear, as in the case of ozone depletion. It is almost as if society is now poised continuously on the edge of its seat, braced for the unexpected "bump" or "vibration" which could throw us completely out of control!

A great deal of uneasiness is also being caused by our not being able to clearly discern where we are headed. It is uncomfortable to realize that you can't go back where you came from and yet know little or nothing about where you are going. Without any significant prophets, authorities, "experts," leaders, and heroes to guide us, we seem to be stumbling into the future. In fact, the future seems to be arriving prematurely, before we are ready for it. Futurist John Platt has said we have gone through a series of "watershed reversals" in the last two decades or so—ecology, detente, dropping birth rates, women's liberation, racial problems, the sexual revolution, religious reform, etc. According to Platt, "It's been like ten Industrial Revolutions and Protestant Reformations all rolled into one—and all taking place in a single generation."[14]

In an attempt to summarize the three points just presented, one might simply ask whether the surprises, "bumps," "vibrations," crises, "watershed reversals," "cloudy crystal ball," etc., that now characterize our highly abnormal age are placing a physical and psychological strain on our "shock absorbers" that they were perhaps never intended to bear. No longer do we seem to know where we are in our "journey," how we got here, and where we are headed! Our age thus seems to be characterized not only by great global physical danger, but also by increasing psychological disorientation and a reduced confidence in the future.

Are there thresholds or even limits to the degree of complexity, change, and disorientation that a society can mentally accomodate? Are there mental limits to growth analogous to the physical limits to growth that the Club of Rome and others proposed in the early 1970's? According to Alvin Toffler, society is already in a state of "future shock." In his 1970 best-selling book of that title, Toffler said, "Future shock is a time phenomenon, a product of the greatly accelerated rate of change in society. It arises from the superimposition of a new culture on an old one. It is a culture shock in one's own society. . . . It is the dizzying disorientation brought on by the premature arrival of the future."

Time and time again, throughout history, when one society has had a cultural change imposed or even forced on it by another, the disorientation produced has proven to be both extensive and long-lasting. For example, even after more than 100 years, the tragic disorientation of the American Indian still remains a major societal problem. An "enemy within" has always been more deceptive, because it is harder to recognize, and thus has always been more difficult to deal with. The largely unrecognized disorientation in society today has resulted from an even more extensive cultural change than that imposed on the Indian—but has been imposed on ourselves, in an even shorter historical period. But, because the enemy is "us," that fact has been largely unrecognized, ignored, or adjusted to.

To bring that point home, one usually doesn't have to look much further than himself. With a little serious thought, does it not become uncomfortably obvious that there has been an almost staggering increase in the acceptance of divorce, promiscuity, premarital sex, illegitimacy, drug use of some type, white-collar crime, dishonesty, mistrust, and greed in just your own circle of influence within the last generation or so? It is actually hard to find a person today who doesn't know of at least one relative, friend, or acquaintance who is routinely and probably openly engaged in some mode(s) of social behavior which societal pressure would have forced behind closed doors only a decade to a generation ago. That person (perhaps yourself!), although probably far more critical of such behavior in the past, is now probably "looking the other way" to at least some extent.

While a relative few believe that the fabric of society is unravelling to the point that society is actually disintegrating, and that humanity may actually be reentering the Dark Ages, most people seem to be simply going about business as usual—and minding their own business! In his book, Alvin Toffler went on to say of future shock, "It may well be the most important disease of tomorrow. . . . The malaise, mass neurosis, irrationality, and free-floating violence already apparent in contemporary life are merely a foretaste of what may lie ahead unless we come to understand and treat this disease." How sick are we? Are we in danger of societal death? How do you tell a people that they are living abnormally in a highly abnormal age, when they believe they are living normally because it is the only age they have ever known? How do you tell a people that they need a prescription when they don't even believe they are sick?

To those who would argue that we are not "sick" and that the previous appraisal is an exaggeration, overly pessimistic, and gloomy, two more comments are in order—comments that can be extracted from what has just been said, and that everyone can agree upon.

First of all, *one can argue about the nature and scope of our current physical and psychological problems, but not whether such problems exist!* It should also be obvious, of course, that those problems will continue to plague us into the foreseeable future (because of their built-in momentum) unless the global exponential growth of population, urbanization, industrialization, resource consumption, and other forces can be stopped. And that seems highly unlikely within the next generation (again because of built-in momentum).

Secondly, both the "optimists" and the "pessimists" now seem to agree that it will take a herculean effort just to meet the crises, challenges, problems, puzzles, paradoxes, dilemmas, and enigmas that are now foreseeable, definable, and perhaps even inescapable in the next generation, let alone the intangibles, surprises, and unexpected problems which will also undoubtedly crop up. The "optimists," of course, believe that our problems can eventually be solved. However, the "pessimists," "doomsday prophets," and "doomsayers" are not so sure. In fact, a very disturbing trend seems to be currently developing in this respect. An increasing volume of literature seems to be appearing in the 1970's that conveys a message of increasing concern, deepening pessimism, and even "giving up" about the future. The following two quotations exemplify this new trend in thought. One can argue about whether the quotations accurately summarize our plight, but who would argue that they are completely unrealistic? In other words, *one can argue about how serious our troubles are, but not whether we're in trouble!* And, increasing numbers of authorities and experts in many fields are expressing their increasing beliefs that we indeed are in an increasing amount of trouble!

> Uncontrollable crises seem to be zeroing in on the peoples of the world. Dwindling food supplies, soaring populations, mass starvation, rampaging inflation, monetary chaos, energy crises, resource competition, political disarray and paralysis, wars and threats of wars, arms races, nuclear proliferation, terrorism, soaring crime, moral decay, weather upsets, pollution, and natural disasters all seem to defy solution by anything short of a new world order.
>
> In the past, many of these problems seemed to be related to each other. Now, they form a perverse, interconnecting web—the "solution" to any one of them often compounding the severity of several others.
>
> —*Plain Truth* magazine, January, 1976

> I speak what I believe to be the truth. It is a dreadful truth, hard to live with. But, if we do not live with it, we shall die by it. . . . Human life is now threatened as never before, not by one but by many perils, each in itself capable of destroying us, but all interrelated, and all coming upon us together. I am one of those scientists who does not see how to bring the human race much beyond the year 2000. . . . Unless the people of this world come together to take control of their lives, to wrest political power from those of its political masters, who are pushing it toward destruction, then we are lost—we, our children, and their children.[15]
>
> —George Wald, 1967 Nobel Laureate in Physiology and Medicine

BIO-SHOCK

To those who now "feel" as well as simply intellectually acknowledge the statement that we do indeed live in the most abnormal, unpredictable, and dangerous period in history, the first and perhaps most important general point to be made about the biological revolution should now be almost obvious. That point will now be prefaced with a very important statement, followed by a rather disturbing question.

The biological revolution, potentially the greatest scientific revolution of all time, is an intangible, a factor which could suddenly and yet drastically change the nature and scope of our already unpredictably dangerous future—with little or no advance warning! Will the rather shaky foundation or structure of society, already under tremendous stress and strain, and undoubtedly under even more in the future, support the added weight of the biological revolution without disintegrating or even collapsing? In other words, in society's increasingly abnormal, unpredictable, and dangerous state, could the biological revolution "tip the balance," "push us over the edge," "be the deciding factor," be "the last straw," or be "the straw that broke the camel's back"?

Are there mental limits to the degree of complexity, confusion, change, flux, and future shock that a society can accomodate, just as some are now saying we are approaching physical limits? If that question is admitted to have *any* legitimacy at all, it should be obvious that the previous question also can not be simply dismissed. For, even if one only accepts the idea that there *may* be limits to societal mental stability, then, at a minimum, the question of whether the biological revolution might carry us past those limits becomes at least debatable. That brings us finally to the first and perhaps the most important point to be made about the biological revolution—the point which this chapter was intended to support. *The biological revolution carries great potential danger for society and perhaps even the threat of societal disintegration and collapse, through the additional, and yet largely unpredictable and unexpected complexity, confusion, change, flux, tension, and future shock that it could add to that which is already foreseeable, and predictable, in the areas of population growth, urbanization, energy consumption, resource depletion, food shortages, etc.* Such a prospect cannot be simply dismissed! It is at least debatable!

BABEL?

Is our abnormal, unpredictable, dangerous age, characterized by almost overwhelming complexity, confusion, change, flux, and future shock, at least the base of a modern Tower of Babel? How solid is that base? Will our modern age support the biological revolution?

CHAPTER 2

BIO-BLESSINGS?

> The optimist claims we live in the best of all worlds. The pessimist fears it's true. —James Branch Cabell

There was a rather interesting cartoon in a December, 1971 issue of *Saturday Review.* It shows a rather tattered, bearded, hippy-type young man who has walked up to the information booth in the middle of a place like Grand Central Station. He then asks the attendant, "Who am I? Where am I going?" The cartoon depicts the attendant as if he is puzzled by the questions, but not altogether dubious.

The cartoon was funny in 1971, but is not so funny today. In a sense, the young man now represents all of us. It is society which seems to be increasingly asking those questions today! And yet, interestingly enough, we are not shopping around for the answers. We seem to be content to let scientists stand in the booth and answer the questions for us. We might as well be honest about it. The scientist is in the booth because since about the middle of the nineteenth century it has been science and technology that have been the leaven that leaven the thinking of our age. In fact, it has been science and technology that have been the major influences and controls in our lives.

Most of the passengers on our "journey" down the superhighway of science and technology have been so intent on enjoying the trip, for so long, that they simply haven't noticed to any great extent the bumps and vibrations that have been developing recently. The trip is finally getting a little boring, but our increasing speed is still quite exhilarating and exciting. For most people, science simply hasn't yet lost that much of its youth and innocence. It may be getting a little tarnished, but its basic luster is still there. The material discussed in chapter one either hasn't registered yet, or else simply hasn't "sunk in." A three-hundred-year string of success stories has created an almost overwhelming tradition of confidence in the abilities and often even the infallibility of scientists and science, and a feeling that very little can really go wrong with our journey. Most of that success has actually occurred in the last 100 years, and is thus still "fresh" in our minds. As Ian Barbour has said,

> In the last hundred years, science has had an impact on almost every aspect of life in the West. Men have been released from back-breaking labor, living standards have risen, and leisure has increased. New drugs, cures of formerly fatal diseases, and the improvement of health standards have more than doubled the average life span in the last century. New products, processes, and machines surround us on every hand, from our electrified homes to our industrialized cities. A trip from New York to San Francisco, which required four months in 1860, takes four hours by jet plane.[1]

Is it any wonder that we seem to believe that it is our destiny to move forward and progress? How many times have *you* heard it said that "You can't stop progress"?

We Americans actually have a built-in psychological resistance or "cushion" that conceals and thus protects us from the idea that anything could go wrong in our journey. First of all, we have enjoyed a historical climb to the "top" and a long stay there, a long tradition of plenty, and a journey through two world wars, a "police action" in Korea, and an undeclared war in Viet Nam, without any of those conflicts touching our shores. Is it any wonder that most of us can't picture anything but an optimistic future? If given a choice, we traditionally now seem to choose the sunnier side of doubt! In fact, at many times today (perhaps too many), we even tend to be giddy super-optimists. Our rose-colored glasses and our soft materialistic lives seem to do an excellent job of shielding us from the harsher aspects of reality.

Secondly, when things do go wrong, we tend to believe that it is not our science which is at fault, but ourselves, for we have been told repeatedly that science is neutral and valueless. It is man that can be good or evil. Science may admittedly build guns, but it is man who pulls the trigger!

Unfortunately, not only do most laymen view science in this way, but most scientists do so as well. Most scientists are essentially optimists, too. Most scientists actually feel that their discoveries and ideas fall like needed rain on society *below,* bringing refreshing showers of material and intellectual blessings. Seldom do you find a scientist who will talk seriously about the limitations and dangers in science, and the possible needs for its regulation, for seldom will you find any scientists who recognize any! When was the last time *you* heard a scientist discuss the limitations and dangers of science, and propose its possible regulation?

And thus it is that most "optimistic" scientists emphasize the benefits and the hope that the biological revolution will bring us rather than its dangers, its promises rather than its possible detrimental implications and consequences, its solutions rather than its problems, and overall, what it will do for us rather than what it will do to us.

As we now begin to examine the possible implications and consequences of the biological revolution in some depth, perhaps we should begin, in all fairness, by taking a look at the "bio-blessings" which the biological revolution will supposedly bring us. The beneficial consequences which will now be discussed will be organized in terms of the five areas of the biological revolution that were used in Part I of the book. It should be stressed that the points listed are intended to only form an overview and are not intended to be complete or all-inclusive. They are simply some of the major beneficial implications and consequences foreseeable by the author at this time. But because no single person can see all of the possible implications and consequences of a discovery either before or even after that discovery is made, the reader will undoubtedly think of at least several more points that he feels should have been on the list.

REPRODUCTION

"Human reproduction by natural means is out of control and must be curbed as soon as possible!" Warnings to that effect have been multiplying and intensifying in the last decade or two. Some of the more dramatic of the warnings were issued in the late 1960's in several books written by Paul Ehrlich. In his 1969 book, *Population, Resources, and Environment,* Ehrlich states his belief that "Spaceship Earth is now filled to capacity or beyond. . . . Population control is absolutely essential if the problems facing mankind are to be solved. . . . The situation is best summarized in the statement 'whatever your cause, it's a lost cause without population control.'"[2]

In the early 1970's a team of researchers, led by Dennis and Donnella Meadows and supported financially by the Club of Rome, attempted to quantify such general warnings in a computer study and resultant book titled *The Limits to Growth.* Like Ehrlich, the team started with the general contention that population growth is the pressure that is causing all of our other problems. However, the computer added the specific warning that unless dramatic progress was made by 2000 in achieving global zero population growth, the world would face global catastrophe and collapse sometime in the 21st century. One can argue about such general predictions, but not about the specific facts upon which they rest.

Until about 1650 A.D., a high death rate acted as a "lid" on world population, holding it to slow, almost linear growth (1, 2, 3, 4, 5, etc.). But with the advent of modern medicine, agriculture, and sanitation, the "lid" came off as the death rate dropped, and world population clearly began to grow exponentially (1, 2, 4, 8, 16, etc.). For example, there were probably only about 100,000 people on the whole earth in 1 million B.C. That population increased 2,500 times to about 250 million people at about the time of Christ—but it took about one million years to do it! Interestingly enough, that means that, even after a million years, there were still only roughly as many people in the whole world at the time of Christ as there are in the United States at the moment. In 1900, there were about 1.5 billion people in the world (about six times as many as there were at the time of Christ) but that increase had only taken about 1,900 years to occur. Today there are about 4 billion people on the earth (almost three times as many as in 1900) but that increase has taken only about 77 years to occur! Another way to demonstrate the *momentum* built into the current exponential growth of world population would be to point out that it took 4.5 billion years (the age of the earth) for the population of the earth to reach 1 billion people in 1830; another 100 years to reach 2 billion in 1930; another 30 years to reach 3 billion in 1960; and just 15 more years to reach 4 billion in 1975. Yet another way might be to point out that if all of man's approximately 3 million years of history could be written in a 3,000 page book, about 1,000 years of history described per page, most of his population growth would have to be described in the last sentence of the book. Still another way would be to point out that if man's history could be measured with a 24-hour clock, most of his population growth has occurred in the last few seconds of the last minute from 11:59 to midnight!

An understanding of the exponential basis of current world population growth brings the specific scope, seriousness, and immediacy of the problem into sharp focus. But, even without that understanding, the need for world population control should be intuitively obvious. One need only ask whether he wants to see even one more doubling of world population, to come to that conclusion. At a current growth rate of "only" about 1.7 percent, that doubling could come before 2020!

Unfortunately, there are only three ways of achieving population control: lowering fertility rates, increasing mortality rates, or migration. But, with increasing restrictions on global migration likely in the future, and continued abhorrence of early death also likely, only one mode of attack actually remains: fertility or birth control.

Such realizations have led to a dramatic increase in global support for the concept of fertility or birth control in the last decade or so. For example, in 1965 less than 20 developing nations had initiated family planning programs. Today more than 50 nations have such programs, many of them far advanced. And, at the World Population Conference held in Bucharest in 1974, 136 nations signed a World Plan of Action calling for a provision of family planning information and means to all individuals and couples desiring them.

In May of 1976, the United States Agency for International Development reported that there were an estimated 165 million couples in the world using birth control in 1975. Of that total, 65 million were estimated to be using sterilization; 55 million, oral contraceptives; 30 million, condoms; and 15 million, IUD's. The numbers are both encouraging and discouraging. Encouraging because an impressive "beginning" has been made in world population control. But also discouraging, for two very important reasons.

First of all, there is a hidden aspect to the birth control problem that can't be seen clearly in the previous figures. With a little reflection, the numbers would seem to indicate that most of the increasing birth control usage to date has been by older, married adults. The global use of birth control among teenagers has simply not kept pace. One can argue, of course, as to whether the use of birth control by teenagers is desirable or undesirable, but not about the seriousness of some recent, global trends among adolescents. Earlier menstruation, increasing global urbanization, and the so-called "sexual revolution" are producing increasing premarital sex and increasing teenage pregnancies almost everywhere in the world.[3] The seriousness of the situation can be seen in the United States, where one might expect the use of birth control by teenagers to be quite extensive. But, such is just not the case. An estimated 30 percent of all U. S. unmarried female teenagers between 15 and 19 now engage in premarital sex, but up to 75 percent of them apparently use no form of birth control.[4] About one-third of the 30 percent, and thus about 1 in 9 of all unmarried, female teenagers overall, are becoming pregnant.

Secondly, it should be obvious that most of the progress in global population control accomplished to date has been in the developed countries. The brakes have been applied least thus far in the underdeveloped countries, where the population acceleration and momentum are actually the greatest. As time goes on, an unfortunate question thus looms larger and larger. Can voluntary fertility control apply enough pressure to sufficiently slow that exponential momentum, or will many countries simply be forced to "slam on the brakes" with compulsory birth controls? India and China seem to be currently on the verge of doing just that. Some believe they have already done it. The final answer to the question could depend largely on whether or not greatly improved fertility controls, in terms of effectiveness and convenience, can be developed and put into use—and how soon.

Overall, it would appear that the development of improved fertility and birth controls would probably have to be considered the greatest immediate, beneficial consequence of the biological revolution.

Continuing on, some possible, additional, specific beneficial consequences in the area of reproduction research will now be briefly outlined. The discussion to follow is not intended to be exhaustive, and the reader is asked to put some "meat on the bones" (and perhaps even add some additional "bones") with his own imagination and predictive ability.

In January of 1976, the Planned Parenthood Federation of America reported that, according to a survey they had conducted, there were 892,000 legal abortions in the United States in 1974, with the death rate from abortion dropping to one-fifth of previous levels. However, they added that an estimated 1.3 to 1.8 million women in all had actually wanted abortions, creating an "abortion gap" of 30 to 50 percent. In May of 1976, the United States Agency for International Development reported that there had been an estimated 40 million abortions in the world in 1975. However, although two-thirds of the women in the world in 1977 live in countries with liberalized abortion laws, compared with only one-third five years ago, the Agency said that many current abortions are still performed illegally, resulting in incalculable suffering and death.

There are many diverse reasons, besides simply legalization, behind the current trend toward increasing abortions in the world. They range from trying to prevent deformed babies being produced by an increasing number of chemical mutagens in the environment to an increasing rejection of "compulsory motherhood." The development of a safe, effective, convenient, and legal chemical abortifacient would undoubtedly "crystallize" the desire for, and availability of, abortion, and thus close the "abortion gap." It would be difficult to predict how many abortions per year such an agent might eventually prompt.

Chemical desire agents could extensively fulfill our age-old search for "instant intimacy" and "life-long honeymoons." The advantages of such agents for the impotent, overly inhibited, unresponsive, and elderly should be obvious. They could also be of great benefit in a marriage between "normal" or "average" people of like or identical ages. For example, many believe that a man's sex drive usually peaks in his 20's and 30's, and a woman's in her 30's and 40's. A chemical desire agent could thus help close any sexual "gap" that might be present in a marriage. It might also prompt more marriages between people of differing ages and perhaps even more intergenerational marriages.

Of course, if a "super" desire agent were developed which not only increased libido but also extended the capability of copulation, stimulated fluid regeneration, and regenerated sex hormones—while staving off fatigue—its additional potential almost staggers the imagination.

The development of chemical anti-desire agents could also be quite desirable. For example, they might be utilized to temper the sex drive for periods of time ranging from days to years, as in illness, pregnancy, service in the Antarctic, extended space journeys, etc. They could also be used to control the sex drive of sex deviates and criminals for long periods of time. Overall, it should be obvious that both chemical desire and anti-desire agents could help control sex crimes and prostitution.

As many as one in every seven couples in the United States suffer from infertility.[5] The development of artificial insemination techniques should thus enable an increasing number of childless parents to have children. The emotional experience of pregnancy and delivery can also be better therapeutically for many such prospective parents than adoption. The development of long-term sperm banks could also provide extended "fatherhood insurance" for those who have undergone sterilization and then change their mind, or who are killed or die unexpectedly. In fact, such "insurance" could even be provided for the whole human race in case of a nuclear war or some other global catastrophe.

Recent census figures for the United States indicate that there are currently slightly more women than men in the population. A slightly higher birth rate for males has apparently been more than balanced by a lower mortality rate for females. However, a recent U. N. report on the role of women indicated that there are currently about 10 million more males than females in the world as a whole. Would the ability to choose the sex of one's offspring result in an overall societal preference for one sex over the other? It is a complex question, but one which could become an important societal issue in years to come.

Looking back through history, there has usually been a preference for boys and patriarchal societies. Based on such age-old, practical reasons as the desire for economic support, old-age security, social standing, and continuity of name and lineage, the preference has ranged from a mild desire in some individuals, countries, and global regions to an obsession in others. Being able to satisfy a preference for boys could now serve another less obvious but extremely important purpose as well. Since a country's fertility rate is dependent on the proportion of females in the population, lowering that proportion by adding more males should lower the fertility rate. Choice of offspring techniques could thus be the most realistic form of long-range birth control in underdeveloped countries. Such techniques would reinforce or satisfy existing attitudes rather than strive to change them, and thus would require little or no propaganda or compulsion.

More recent trends indicate a historical shift in preference toward females is actually underway. Women's liberation and an increasingly matriarchal society could reinforce such a shift in preference. Rejection of our brutal, war-like age could also lead society back to valuing the beauty, love, sensuality, compassion, and loyalty that are usually more associated with women. And it is becoming increasingly clear that, biologically speaking, females are usually stronger and healthier than males. It is now known, for example, that there are more than 50 sex-linked, hereditary diseases (e.g. hemophilia) that affect males exclusively. Parents thus afflicted can avoid passing these diseases on to their children only by having girls.

The development of artificial inovulation, the artificial womb, and "baby factories" will augment many of the benefits just described, particularly those for artificial insemination. They will also help remove the fear, inconvenience, and danger of childbirth for many parents, and remove a potent set of neuroses, as childbearing is completely removed from sex. The embryo and fetus will also profit, being protected and safe from the possible ills, emotional shocks, malnutrition, drug use, accidents, death, or murder of a real mother. In fact, the artificial womb could even protect the embryo from the illnesses and increasing number of environmental chemicals and agents being found capable of producing genetic damage to the unborn.

In his 1975 book, *Birth Without Violence,* Frederick Le Boyer contends that most children also meet with traumatic shock at birth—blinding lights, noise, cold air, harsh fabrics, being snapped into an erect position, being held aloft by one foot, being slapped on the rear and forced to scream, etc. His recommendations for alterations in the birth environment and process could all be readily implemented with the artificial womb and "baby factory."

Perhaps the most astounding possibility yet advanced for using artificial inovulation and "baby factory" techniques was proposed about 15 years ago by the late Hermann J. Muller, a pioneer geneticist and Nobel prizewinner. In Dr. Muller's proposal, all children would be sterilized at birth, after placing a sample of their sperm or eggs in deep freeze. After their death, records of each individual's accomplishments and characteristics during life would be reviewed. If it was decided that a certain individual was a "desirable" person while alive, his or her frozen sex cells would be combined with those of another "desirable" person, in order to produce one or more "desirable" babies with artificial inovulation or "baby factory" techniques. The sex cells of "undesirable" people would simply be discarded. There has been a rekindling of interest in Muller's proposal recently, for it now appears that the techniques needed to carry out his plan are either available or close at hand.[6] Such a "super race" could probably be started within the next 25 years—or less!

The immediate, practical benefits of being able to duplicate or clone "desirable" species of wildlife, and perhaps even "desirable" people eventually, are also so staggering in their possible scope and potential that they are essentially limited only by the extent of one's imagination. As just one example, some have envisioned a "brown revolution" of cloned livestock and other animals that might help resolve the world's food problems.

There may also be some longer-range benefits that are even more mind-boggling. One of the more impressive possibilities involves the largely unrecognized, but looming collision between man's deep need (probably genetic) for the beauty and diversity of nature, and "the waning wild." Man has been a predominantly "rural animal" for more than 99.99 percent of his existence, being a predominantly "urban animal" for only the last 150 years or so. But, as his cities grow, the beauty and diversity of nature is disappearing—as can be seen in the following, sobering facts. There are approximately 4,300 species of mammals alive today. But several hundred—possibly 1 in 10—are currently in serious danger of extinction.[7] Overall, more than 1,000 species of birds and animals are now threatened with extinction, some imminently, some within a decade or so. More than 400 are native to the United States.[8] Wildlife is now disappearing in the world at a rate of one species or sub-species per year, and the rate is increasing rapidly.[9] In April of 1976, the U. S. Department of the Interior proposed the addition of many more species to the list of those now endangered or threatened. Their proposal included 27 species of primates, raising the total to 55, or 35 percent of the world's monkeys and apes. The list now includes the chimpanzee, the most intelligent of all animals, and the orangutang, perhaps the most docile and man-like of all the primates. There is also a less-publicized component to the list of endangered wildlife—plants and insects. In a 1975 report to Congress, the Smithsonian Institution's National Museum of Natural History stated that nearly 3,200 kinds of higher plants native to the United States are endangered, threatened, or recently extinct—over 14 percent of the nation's floral heritage! In the last few years, sources too numerous to mention have been issuing warnings that literally hundreds of plants and insect species should be added to the list. Even the common honeybee seems to be in increasing danger—another potential environmental catastrophy in the making.[10]

By combining long-term, cloning "cell-banks" with the artificial womb, artificial inovulation, and artificial insemination techniques, not only could domesticated wildlife be multiplied on a temporary basis, but the extinction of rare and endangered species (and individual specimens) might be prevented. In other words, evolution and natural selection could be bypassed to some degree by "freezing" certain desired, distinctive, historical continuities in place.

The historical continuity and evolutionary distinctiveness of individuals might also be established one day with cloning. Today, mankind is immortal, but tomorrow *individual persons* could enjoy the same prospect! An individual could be duplicated upon his death from some of his cells placed in a "cell bank." Or, it seems likely that many parents would prefer children that were genetic copies of either or both of them. This would certainly be infinitely preferable to the use of anonymous donors by those using artificial insemination techniques today. The possible consequences of the multiplicative immortality of certain humans are almost impossible to predict. But one prospect is particularly intriguing. Today we invariably seem to relive history because we seldom seem to learn our historical lessons. What would happen if our leaders learned their historical lessons from accumulated, long-term experience, hereditarily passed from one clone to another, rather than merely from books? Would the words "collective wisdom," agreement, consensus, etc. take on new meanings? Might not a whole new age of wisdom arise? If not, at least certain nations could clone top leaders, scientists, soldiers, etc. to offset the use of similar strategies by more tyrannical nations!

It may even be necessary (as well as desirable) to clone people some day. As Dr. Joseph Fletcher, professor of medical ethics at the University of Virginia has said, "It is entirely possible, given our current progressive pollution of the human gene pool through uncontrolled sexual reproduction, that we might have to replicate healthy people to compensate for the spread of genetic disease."[11]

In summary, reproduction research could simultaneously solve the world's population problems and also free women from the burden of unwanted, unloved children. But, it could also provide us with greater mental health and more enjoyment of sex. Elimination of the fear of pregnancy alone should allow people to better center their lives around their own desires. Sex could thus become a more spontaneous, "natural function," leading to a disappearance of inhibitions caused by the idea that sex is "bad" or "evil," and this too could lead to greater mental health and enjoyment. If God desires that man also be "saved" sexually, his sexual salvation may be near.

Sociologist Leo Davids has been more blunt in similar predictions.[12] He believes that within two decades, two myths will have largely disappeared: the myths of "romantic love" and "parenthood is fun." He also believes illegitimacy and venereal disease will have disappeared, that there will be several kinds of marriage to choose from, and that prospective parents will have to obtain licenses and undergo extensive training before having children!

PHYSICAL MODIFICATION

The increased use of transplanted, artificial, and regenerated body parts will help minimize the effects of disabling injuries, greatly alleviate the physical, mental, and financial pain, suffering, and incapacitation caused by aging, and thus result in overall increases in our life-span. There will also be more subtle benefits from the use of such techniques. One example might be the chance for suicidal persons to donate organs, possibly a way to conquer their self-destruction tendencies, by increasing their self-esteem, decreasing their egoism, relieving their guilt, and getting them more involved in society.[13]

In the 1970's, the rapid (and often largely unexpected) development of such genetic techniques as gene mapping, recombinant DNA, etc. (as outlined earlier in Part I), has resulted in an almost overnight injection of "believability" into the prospects for genetic engineering. For example, many researchers are now much more optimistic than they were just several years ago that negative eugenics may eventually be used to solve two tremendous societal physical modification problems.

First of all, there is the obvious need for a cure for cancer, particularly for the one in every five people who will die from the disease. Unfortunately, the cancer threat may be increasing, if a recent theory about the origins of the disease turns out to be correct.[14] The theory combines the virus and genetic theories currently in vogue, and adds an interesting "twist." It contends that cancer-causing viruses have been passed on genetically from the evolutionary past, but remain unexpressed until somehow activated by radiation, carcinogenic chemicals, or other factors. Chemicals are apparently far and away the more important of the "activators" today. The World Health Organization (WHO) estimates that 60 to 90 percent of all cancer is chemically-induced.[15]

There are about 250,000 chemicals already in daily use. Only a handful of the approximately 30,000 pesticides registered (containing over 1,200 active ingredients) have been thoroughly tested for their safety, however, and, of the over 3,000 food additives registered, only about 700 are recognized as being "safe." Overall, only about 6,000 chemicals have been thoroughly tested for their safety, and about one-third of those have been found to be toxic, carcinogenic, and/or mutagenic![16] In April of 1975, the National Institute for Occupational Safety and Health announced it would investigate the occupational hazards of about 1,500 chemicals now suspected to be carcinogens or tumorigens.[17] The list seems to be growing almost daily, with even vinyl chloride, a component of one of our most extensively used plastics now a suspected carcinogen—and mutagen![18] There are also about 300,000 new chemicals added yearly to the list of over 2 million already known, with about 700 new chemicals placed in production yearly.

If a growing chemical threat is actually creating an increasing immediate and long-term incidence of cancer, as many now believe, there is obviously a growing need for negative eugenics. Cancer can only be controlled by restricting the use of chemicals, and even that may now be increasingly difficult. In order to actually cure cancer, it will probably be necessary to dip into the genes of viruses, healthy cells, or both.

Eliminating genetic diseases such as Down's syndrome (mongolism) could be the second great victory for negative eugenics. In fact, there are at least 1,500 distinquishable diseases now known to be genetically determined. Many of them are unfortunately largely confined to specific ethnic or racial groups. Sickle-cell anemia, for example, occurs chiefly in blacks, and cystic fibrosis mainly in whites. Six of the most devastating genetic illnesses are found primarily in Jews: Tay-Sach's disease, Niemann-Pick disease, Gaucher's disease, torsion dystonia, familial dysautonomia, and Bloom's syndrome. Overall, about 1 in every 20 babies born in the United States each year has some sort of genetic defect. About 1 in every 200 has a defect leading to mental retardation or physical disability, twisting the joy of parenthood into lingering tragedy. There may be as many as 15 million people in the United States with congenital diseases serious enough to affect their daily lives! There is also a financial burden associated with genetic disease, probably exceeding several billions of dollars per year for mongolism alone.

Many scientists now believe that the human gene pool is being increasingly polluted with defective genes. One can argue about the extent to which those with genetic diseases are contributing to the problem.[19] However, it cannot be argued that there is an increasing use and abuse of a growing number of diverse chemical mutagens in everyday life. There also seem to be an increasing number of "accidents" with those chemicals, ranging from the thalidomide disaster of the 1960's to the extremely serious release of dioxin (TCDD) in Italy in 1976.[20] There are other causes as well, such as the use of chemical agents in Viet Nam that can cause birth defects, the fact that an increasing number of teenage pregnancies are resulting in increased birth defects, etc.[21] There is even a small but increasing body of evidence to indicate that "electromagnetic pollution" from radio and television transmitters, power lines, "leaking" microwave ovens, etc. may be causing genetic damage.[22] Overall, if humanity actually is slowly but irreversibly sinking into an increasingly polluted gene pool, only negative eugenics may be able to eventually bail it out.

The development of gene mapping, recombinant DNA, and other genetic techniques in the 1970's has also scrambled previous predictions that positive eugenics would be utilized only far in the future, if ever. While it is still probably true that the development of "superman" is just not possible before the next century, many more practical projects are already "on the drawing board." One of the more exciting general blueprints to emerge thus far can be found in the recent, parallel predictions of Joshua Lederberg and Freeman Dyson. Lederberg believes the plasmid technique "will lead to a technology of untold importance"—a cheap means of producing antibiotics and other medicines, essential nutrients, industrial chemicals, and even human proteins.[23] Dyson believes we may actually be on the verge of a "second industrial revolution" that could "undo the results of the first one."[24] It might be possible, for example, to produce bacterial organisms with the plasmid technique that would allow recycling animal, agricultural, and mining wastes, sewage, refuse, manufacturing by-products, and effluents. In fact, a very important first step in this regard has already been taken.

In 1975, Ananada M. Chakrabarty of General Electric announced the development of an oil-digesting microbe comprised of recombinant genetic material from four different natural strains of *Pseudomonas.* The hybrid organism can digest two-thirds of the hydrocarbons present in crude oil spills.[25] Work is also underway on a second organism which can digest those portions of crude oil which the first strain cannot. Chakrabarty believes that similar bacteria can also be produced which will produce protein from crude oil and allow man to digest the cellulose found in wood, weeds, grass, etc..[26] Actually, there are so many additional, specific uses for positive eugenics now being envisioned and planned that only a few more diverse examples will be cited, with no elaboration of each. It might be possible, for example, to make insulin in commercial quantities by transferring insulin genes to rapidly-reproducing bacteria,[27] produce rapid, super-weight gain in animals and fowl, change one's skin color, produce "made-to-order" astronauts,[28] elucidate the mechanisms of evolution,[29] come to an understanding of the possible nature of life outside the earth, or even alter the instincts and biological rhythms of man and other forms of life.[30] Techniques to accomplish the latter could lead to the eventual alteration and even control of menstrual cycles, sleep cycles, reproductive cycles, migration, and cell proliferation and division.

Finally, with regard to agriculture, it may be possible, among other things, to increase insect, blight, and disease resistance in various crops, alter the reproductive genes of various insects in order to control their reproductive cycles,[31] and transfer nitrogen-fixing genes to bacteria that live next to the roots of corn, wheat, etc., in order to at least partially eliminate the need for fertilizers.[32] In fact, an understanding of the genetic basis of the chlorophyll process might someday even make possible a biomimetic technology that would convert sunlight directly to chemical energy.[33] Such advances could come none too soon. Some believe that the art of genetic improvement as it has been applied to modern agriculture is resulting simultaneously in increasing genetic uniformity ("monocultures") and decreasing natural genetic diversity ("genetic erosion").[34] Such researchers also claim that the 1970 "Southern corn blight" was a stunning demonstration of the genetic vulnerability of many currently grown crops, but a mere warning of the global agricultural catastrophe which may lie ahead. If they are right, the *science* of genetic improvement—positive eugenics, chimeras, and synthetic plants and animals—could well become a critical necessity, as well as simply an intellectual endeavor! We may have to speed up the time scale that James F. Daniclli foresaw in 1970 when he said, "In biology we are moving from an age of analysis into an age of synthesis. Within a century, we shall probably be able to synthesize artificially any biological system or entity."[35]

MENTAL MODIFICATION

There can be no argument about the need for global behavior modification. In late 1975, the U.N. World Health Organization (WHO) reported that there is a "virtual epidemic" of behavioral disorders currently sweeping the world.[36] Increasing global alcoholism, drug dependence, and heavy smoking were all said to be symptoms of increasingly sick consciences, impoverished souls, and alienation from society.

Such behavioral disorders are producing many serious consequences. One of them is increasing global suicide. Dr. Anthony May of the WHO has estimated that more than 1,000 people commit suicide every day in the world, with more than 10 times as many attempting it.[37] Suicide is now the fourth leading cause of death in most European countries. Jacques Charon, author of *Death and Western Thought*, has estimated that between six and seven million Americans have attempted suicide. Suicide is now the second leading cause of death among the young in the United States!

The most serious consequence of behavioral disorders, however, is aggression. Man has probably been at peace less than 10 percent of his existence, despite thousands of treaties. Neither has "societal craziness" shown any signs of diminishing. There have been over 50 undeclared, major "struggles" in the world since World War II. Most of them have been in the underdeveloped countries—although not all. The undeclared "wars" in Korea and Viet Nam rivaled the two world wars in scope, intensity, and casualties.

The extensive development of electrical and/or chemical behavior modification techniques could greatly alleviate pain and suffering, control suicidal tendencies by alleviating depression and despondency, alleviate or eliminate aggression and many other anti-social characteristics, help the mentally and criminally ill, and decrease senility and boredom. The overall result should be increased happiness, enjoyment, and satisfaction in life. In fact, the latter three states of mind may in themselves be directly inducible with "happiness and pleasure wiring" or the use of an appropriate "pill."

The development of intelligence and/or memory modification techniques could also add further possibilities for societal improvement which are too extensive to list, except for a few random, generalized statements. For example, there are more than 50 million people in the United States with IQ's under 90.[38] And, if intelligence enhancement techniques can also be applied to primates, the servant and worker apes depicted so vividly in the 1972 movie, *Conquest of the Planet of the Apes,* might not remain science fiction!

Aggression may also be amenable to a direct electrical and/or chemical attack. On September 4, 1971, Kenneth B. Clark, in a presidential address before the American Psychological Association, advocated the development of a "peace pill" which would prevent those in power from abusing it. He contended that man does not have time to lessen aggression and prevent war by environmental improvement. In response to those who have since asked "who will control the controller?" Clark has asked in return "who controls them today?"[39]

In some ways, it might be argued that the application of negative eugenics to mental problems could result in societal benefits at least as great as those resulting from curing cancer or genetic diseases. For example, some experts believe that as many as one in every four adults in the United States may be in need of psychiatric help. Although many of the troubled unfortunately remain troubled, others are luckily directed to help by ministers, social workers, the police, etc., and many more voluntarily seek help on their own. About 350,000 adults are currently admitted to mental hospitals per year. The situation with regard to troubled children is far less encouraging, however.

According to some estimates, there are at least 1.4 million children in the United States under the age of 18 with emotional problems severe enough to warrant urgent attention. As supportive evidence for that fact, it might be pointed out that the number of children receiving treatment for emotional problems in institutions and outpatient facilities has risen nearly 60% in the last seven years, from 486,000 to 770,000. As many as 10 million more children are plagued by neurotic symptoms severe enough to require psychiatric help of some kind, if they are ever to reach their full potential. It would be hard to pinpoint and accurately estimate the number of children who simply don't function properly in society. Overall, some believe that 10% or more of the nation's children are destined to develop some form of mental and emotional disability.[40]

There are those, of course, who believe that negative eugenics will never be of any significant help in solving mental and emotional problems. They point out, justifiably so, that it has not yet been determined to what extent such problems are even genetically caused. And, they point out that even if such is found to be the case, the identification and correction of the *multiple* genes undoubtedly involved will be far more difficult than with the individual genes involved in physical defects. Overall, it does seem safe to predict that physical negative eugenics will be attempted first. However, it also seems likely, that, because of the need just outlined, any degree of success at all in the physical area could create tremendous, mushrooming pressure to attempt mental negative eugenics as well.

The development of mental negative eugenics might also shed considerable light on two questions which have been receiving increasing attention in the last decade or so. In the late 1970's, these questions passed rather quietly beyond the point of simple intellectual interest into the realm of societal significance. A small, but increasing number of researchers are beginning to urge that the questions be resolved.

The first question is simply whether or not our aggressiveness is inherited, acquired, or some combination of both. If aggression is at least partially genetic, it is conceivable that it might be as treatable or curable, just like other mental problems, through negative eugenics. The idea sounds rather farfetched at first glance. However, the Qolla, an Andean subculture inhabiting the area around Lake Titicaca, may be living proof that it may actually be possible. The Qolla have a homicide rate higher than any nation on earth and have been described as being the world's meanest and most unlikable people. Their problem may be genetic. As early as 1947 researchers suggested there might be a link between hypoglycemia (an abnormally low blood glucose level) and aggressivity. After living with the Qolla for five years, ethnographer Ralph Bolton has hypothesized that their aggressiveness is due to hypoglycemia.[41] It might also be pointed out that Edward O. Wilson's 1975 book, *Sociobiology:*

The New Synthesis, has injected a heightened interest into the question of the possible genetic basis for aggression and behavior.[42] Wilson contends that there is such a basis, and that it is exhibited most plainly in the insect world where "kin selection" and "altruism" are expressed fanatically.

The second question is still being discussed only in whispers for the most part. There is a rather heated argument currently taking place behind the scenes as to whether or not some races are genetically superior to others. To date, the argument has largely centered around the issue of intelligence. Nobel Laureate William Shockley, Stanford physicist and co-inventor of the transistor, has probably been the chief spokesman for those on the pro side of the controversy.[43] He believes that intelligence is over 80 percent hereditary and only about 15 percent acquired. He then uses that belief in support of the contention that blacks consistently score lower than whites on IQ tests for genetic reasons. Shockly and others maintain that it is time the question of possible racial superiority is resolved, one way or the other.

The development of techniques for producing man-computer chimeras, man-machine chimeras, brain transplants, and disembodied brains (cyborgs) could produce beneficial consequences almost too vast and far-reaching to even attempt to contemplate. In summary, the techniques would enable man to improve his thinking ability and assure the continued availability of great thinkers and other "desirable" people. Freeing the brain from the body and/or connecting it to computers and machines could also provide whole new horizons of sensations and thought.

Overall, as man moves to a new mental "oneness," through mental modification, he could reach new heights of tolerance, understanding, and cooperation.

THE PROLONGMENT OF LIFE

Success in prolongment of life research would produce four general benefits whose overall value should be largely obvious. First of all, the fact that all too often "death leaves life's work undone" probably justifies aging research in itself. But, secondly, there would be incalculable value in alleviating or curing gerontophobia (fear of aging), probably humanity's most prevalent disease. Thirdly, the same could be said for lessening or even finally ending the age-old quest for the "fountain of youth" and the age-old dream of immortality. Hopefully, the historically endless energies and monies that gerontophobia, the quest, and the dream have consumed might then be directed into more useful channels such as finishing "life's work." And, lastly, increasing the proportion of people in the maturing, mellowing, "settling down" period of life could lead to a more stable, more enjoyable society.

The four techniques discussed earlier for possibly prolonging life will also provide some specific benefits unique in themselves. For example, the control or elimination of disease will enable people to live longer, with less pain, suffering, and disability. The development of freezing techniques and/or suspended animation chemicals could open the door to both extended space travel, or, in a sense, time travel here on the earth. Man has dreamed of accomplishing both for centuries. The development of an anti-aging drug could help equalize the current approximately eight year gap in the life expectancies of men and women. And overall, the development of any or all four of the techniques could shed considerable light on all of the other research aspects of the biological revolution.

THE CREATION OF LIFE

The specific benefits of success in creation of life research will undoubtedly be far too numerous and diverse to be completely foreseeable at this point. However, in general, it might be pointed out that such research will once again augment and illuminate other research aspects of the biological revolution. It could also produce new forms of life—some of them a blend of human characteristics with those of other species. It could also produce a greatly heightened interest in the possibility of extraterrestrial life. Overall, successful creation of life research could simply be humanity's greatest achievement. With the creation of life, we could well answer perhaps our oldest questions: how, when, where, and why did we originate?

IN SUMMARY

It should be obvious that if only a fraction of the beneficial consequences just described reach society, there would indeed be ahead, as one chemical company has put it, "Better Things for Better Living—Through Chemistry." In fact, some of the developments which the biological revolution could produce, as, for example, improved birth control, are actually absolutely essential if humanity is to continue to live viably on this planet. However, won't the extensive, global use of birth control also carry a "price," and perhaps create serious problems as well as provide beneficial solutions? And to what extent would even improved birth control add to the societal complexity, change, flux, and future shock that now seem to so prominently characterize our age? These are the types of questions that are receiving very little emphasis today, and yet almost beg to be asked—and answered!

To what extent will bio-blessings prove to be mixed blessings? The intimation that there may be trouble as well as blessings among the foreseeable beneficial consequences of the biological revolution is at least somewhat disquieting. Intuitively, one almost begins to envision the hazy, indistinct outline of a potential Bio-Babel beginning to take form on the shaky foundation discussed in chapter one, and that thought is indefinably disturbing!

CHAPTER THREE

BIO-DANGER

> To every man is given the key to the gates of heaven: the same key unlocks the gates of hell. —Buddhist Saying

On November 1, 1914, a German fleet met and defeated a British fleet at Coronel off the coast of Chile. The remnants of the British fleet escaped to Falkland Island, were reinforced with additional ships, and again met the German fleet, which was destroyed except for one lone ship, which escaped.

Many believed that these two battles would end World War I while it was still in its infancy, for the nitrate deposits of Chile had historically been the world's primary source of nitrates for explosives. By denying Germany access to the sea lanes to Chile, she was also obviously denied a valuable asset, and perhaps even a prerequisite, in her plans of carrying on a large-scale war for an extended period of time. But it just didn't turn out that way. By twisting a historical coincidence to her advantage, Germany was able to fight on.

The coincidence developed as the result of a growing imbalance in the nitrogen cycle in the late 1800's. In the absence of man, nature continuously adds nitrates to the soil to replace those removed through growing vegetation, through the decay of plant and animal protein, the action of nitrogen-fixing bacteria in the soil, and the action of lightning on atmospheric nitrogen. However, as man grows crops in addition to or in place of natural vegetation, he upsets the cycle and must compensate by adding fertilizers to restore the balance. Unfortunately, toward the end of the 1800's, man's ability to produce natural fertilizers such as manure, compost, bones, and nitrates was beginning to lag behind his increased growth of crops. About the turn of the century, Sir William Crookes, a famous nineteenth century scientist, was one of the first to sound the alarm. He and others began to increasingly call for the development of techniques to make synthetic fertilizers.

And thus the coincidence unfolded, for the first such techniques were developed in Germany. In 1902, Carl von Linde perfected a method (the Linde Process) for isolating nitrogen from the atmosphere. A method for producing hydrogen from coal and steam had been in use since about 1860. In 1909, Fritz Haber developed a method of nitrogen-fixation (the Haber Process) in which nitrogen and hydrogen are forced to form ammonia with high temperatures and pressures. Earlier, in 1907, Wilhelm Ostwald, another German chemist, had developed a method (the Ostwald Process) for converting ammonia directly into nitric acid by burning the ammonia with air over a hot catalyst.

The desired end product of the three processes is nitric acid, which can be combined with potassium carbonate (potash) and ammonia to produce potassium nitrate and ammonium nitrate—both excellent explosives as well as fertilizers. In fact, nitric acid can also be used to make other explosives as well, such as TNT, nitroglycerine, nitrocellulose, etc. And thus it was that a historical coincidence was twisted by the irascibility of human nature. The four years of war that followed claimed the lives of over four million people!

Many would point out, of course, that the three aforementioned processes have also been used since the early 1900's to make synthetic fertilizers—the purpose for which they were actually originally intended. And, it is also true that many people's lives have been sustained and even saved through the use of synthetic fertilizers. In fact, the three processes currently make possible over one-third of all the food grown in the world. But it could also be countered that the world is apparently becoming increasingly addicted to the use of synthetic fertilizers and that the addiction is beginning to create serious environmental problems. Although it is an uncomfortable thought, it could also be pointed out that, in sustaining and saving the lives of so many people who would have otherwise died earlier in life, synthetic fertilizers have helped produce a global population explosion in which the world's population has more than doubled in just three generations, from 1.5 billion people in 1900 to over 4 billion today.

The synthetic fertilizer story is only one of many which could be cited to dramatically illustrate the paradoxical, double-edged, sword-like nature of science and technology today. Almost every discovery now made seems to carry the same promise of paradoxically doing both detrimental things to us as well as beneficial things for us, of doing us great harm as well as great good. We examined many of the foreseeable beneficial consequences of the biological revolution in the last chapter. Turning the sword, so to speak, it is time to now examine its negative cutting edge.

Once again, as in the last chapter, it should be stressed that the following discussion consists of only those consequences currently foreseen and considered to be potential future societal problems by the author. The list should not be considered to be complete or all-inclusive, but rather simply a stimulative overview. The reader may thus wish to again make additions or deletions in the list. One can admittedly argue about where each consequence might be on a scale ranging from probable to improbable (or even farfetched!), and also about *when* the consequence might emerge as a societal problem, even if its emergence is considered to be probable. The consequences listed will thus be largely phrased in the form of questions, since at this point in history, that is actually all they are. There are no definite answers as yet available to any of them!

REPRODUCTION

Fertility (Birth) Control

1. Are there times in history when the rights of society supercede or take precedence over the rights of an individual? In this unique historical age of critical global overpopulation and extensive adoption possibilities, do individuals really have a right to reproduce? To a limited or unlimited extent? If to only a limited extent, does society have a right to set the limits? Is it right in our age of limited resources and unlimited needs to take a sizable fraction of the finances and efforts of medicine and direct them toward the problem of infertility, as if it were an illness?
2. Should there be a balance struck between the "quality" and the "quantity" of life? What would be an optimum quality of life and quantity of people for the United States? The world? Who should decide?
3. If the use of birth control drugs is to be greatly increased on a global scale, a great deal of *in situ* testing may be necessary in the underdeveloped countries. For example, primary questions such as the possible correlation between drug safety and body size, diet, disease, malnutrition, etc., will have to be resolved. Such primary questions will raise secondary questions such as how much testing should be done, where, with what standards, with how much medical supervision, with how much protection and education of the subjects, etc. Will a trend toward more caution and emphasis on risks in the use of such drugs in the developed countries prove to be a roadblock in attempts to greatly increase their use in the underdeveloped countries, or will a double standard of safety simply arise?
4. Publicized side effects such as the clotting risk can be confusing and even frightening in the underdeveloped countries. Could such publicity in the developed countries actually reverse attempts at birth control in the underdeveloped countries?

5. By lowering the acidity of the vaginal area and thus decreasing its germ-killing capacity, while meanwhile decreasing the use of condoms, is it possible that the pill is actually promoting venereal disease?
6. Some short-range effects of the pill have already been documented (e.g. clotting). Any possible long-range effects will only show up after a long-range period of time has passed. Will there be any?
7. Some have argued that the so-called "sexual revolution" is a fabrication, and that sexual expression and freedom are simply less concealed today than they used to be. But recent polls, trends, and events do not seem to bear that contention out. In fact, it can be clearly documented that gonorrhea and syphilis are currently increasing, to the point of now being our number one and number three infectious diseases respectively in society. These two diseases, and 12 others related to sexual freedom have been called "the Silent Epidemic." Some have said that the existence of the epidemic is evidence that "free sex is not so free." It has also been stated that women's liberation is producing more female sexual aggressiveness. Overall, it would seem that sexual expression and freedom *is* on the increase in society today. Is that a healthy trend? What is causing it? Where will it end? Is the increased availability of birth control agents, and especially the pill, a major factor? Only a fraction of the nation's girls and women use the pill today. What new and/or enhanced attitudes toward sex, love, and marriage will result from the use of far more sophisticated and easier to use birth control agents than those now available? Should such agents be made available to those under 18 or even younger?
8. In December of 1976, the first Interhemispheric Conference on Adolescent Fertility stated that teenage pregnancy worldwide had become an "epidemic."[1] The World Health Organization reported earlier, in 1975, that venereal disease is rapidly becoming a global epidemic. Is the sexual revolution a desirable or undesirable trend for the world? To what extent is it being prompted by current means of birth control? To what extent could the trend be accelerated by more sophisticated and easier to use birth control agents? Where could such a trend eventually lead?
9. To what extent is birth control, and the "instant sex" which it allows, synergistic with other societal trends? For example, with societal mobility to produce societal "rootlessness"?
10. Is voluntary birth control by individuals a step toward compulsory control by society? Should sterilization be forced on certain people? For example, the genetically ill, the retarded, the criminally insane? How about society as a whole? Should methods be developed for controlling the fertility of an entire population if the voluntary use of

conventional methods fails to stem population growth? Although many technological problems stand in the way, some advocate the development and addition of a temporary sterilant to water supplies and/or staple foods. If an antidote was available, who should have access to it? Who should decide that?

Abortion

In a 7–2 decision on January 22, 1973, the U. S. Supreme Court struck down all existing state laws preventing abortion in the first six months of pregnancy. The decision stated the question of abortion can be entirely decided by only a woman and her doctor in the first 24 weeks of pregnancy and that the state can only proscribe abortion in the last 12 weeks. The decision has generated continual controversy ever since it was rendered. In fact, there was even an anti-abortion candidate for President in the 1976 elections.

1. Is there an overall balance point between the quality and the quantity of life in a joint consideration of an unborn fetus, its mother and father, an unwanted child, and society at large? Where is that point?
2. At what point in its early history as a living organism does a human being become a human being? Does a person become a person at the moment of conception, at the transition of an embryo to a fetus (about three months into pregnancy), at the point of heart viability (about 24 weeks into pregnancy), at the point of brain viability and the first signs of human intelligence (about 28–32 weeks into pregnancy), or at birth? Of these five events, only the moment of birth can be determined with any real accuracy. But the answer to the question is extremely important, for it apparently also defines the point at which a person acquires the legal rights that grant protection from death or disposal. In the U.S. Constitution, what do the words "right to life" mean? If that "right" is acquired before birth, at what point in pregnancy, if any, does it become legal to punish a fetus with death for the sins of its mother or father, as in the case of adultery or rape? At what point, if any, does abortion become murder? Is it unethical to destroy a fetus even if it is legal to do so? Self-preservation in the face of the threat of extinction is supposedly the only ethical grounds on which to take another's life. But while the extinction of the human race is possible today, few would argue that it is actually imminent.

On February 15, 1975, Dr. Kenneth C. Edelin, a Boston City Hospital obstetrician, was convicted of manslaughter in the death of a 20 to 24-week-old fetus he aborted at that hospital on October 3, 1973, at the request of a 17-year-old girl. Edelin was later sentenced to a year of probation. However, in late December of 1976, the Massachusetts Supreme Judicial Court reversed the conviction and the sentence of the lower court. One of the two courts obviously made a mistake. Which one?

3. If abortion is immoral and unethical, is contraception also immoral and unethical, since their ultimate purposes, motives, and intent are the same?
4. Abortion is now the leading cause of death in the United States. And yet, in January of 1976, the Planned Parenthood Federation of America announced publicly that, according to a nationwide survey they had conducted, 892,000 abortions were performed in 1974, but an estimated 1.3 to 1.8 million woman had actually wanted them. To what extent does an "abortion gap" produce unwanted children? Is not a child's first right above all others the right to be wanted? Is it actually unethical to bring an unwanted child into the world? Do parents have a responsibility to sacrifice at least some of their hoped-for-future for an unwanted child? Does society have a responsibility to sacrifice for and care for unwanted children the way it takes care of others?
5. Should abortions be made even easier to obtain than they are today? For example, are increasing crime, the use of an increasing number of chemical mutagens, etc., gradually increasing the need for abortion? Or, are such factors prompting people to "pull the trigger" on abortion? Is it already too easy to obtain abortions today?
6. In August of 1975, the 13-month ban on government-funded fetal research was lifted, just as opponents of abortion predicted it would be, after the earlier Supreme Court decision. In removing the ban, Caspar W. Weinberger, Secretary of Health, Education, and Welfare, also revealed new federal guidelines which permitted research studies on fetuses that are still in the womb of the mother, as well as fetuses delivered in mid-term abortions, during the moments the fetus is still alive. No experiment is allowed, however, which artifically keeps a fetus alive unless the study is aimed at ways of saving premature babies. Is such research ethical and justifiable? Or, is it immoral and unethical because the fetus cannot give its consent? Would banning medical research on living fetuses deny freedom from disease to millions of future children, as many scientists claim?
7. Aborted fetuses are sometimes live-born and live for an extended period of time afterward, especially in abortions that occur late in pregnancies. In such cases does the doctor then have a responsibility to save a fetus which was previously unwanted? Is that not a rather paradoxical situation?
8. The predictive value of amniocentesis (the sampling of the amniotic fluid to study the fetus' chromosomes) is being increasingly coupled

with abortion to reject fetuses with such genetic diseases as mongolism. Might not such techniques gradually come to be used to permit other rejective choices as well (e.g. sex selection) which such chromosomal studies can also reveal?

9. Various polls and studies have shown that, when abortion is legal and accepted culturally, the vast majority of women having abortions show little or no evidence of guilt feelings and psychological problems. Where is the balance point between a woman's rights to privacy, peace of mind, reproductive freedom, and making a moral and ethical choice consistent with her conscience and religious beliefs, and a fetus' right of life? Does the right to life supercede a woman's rights? Or, is a law prohibiting abortion, and thus dictating compulsory motherhood, as indefensible and unethical as one would be advocating compulsory abortion? Those opposed to abortion wish to ban it for everyone! Would this restrict the religious freedom of others and their right to make a free moral choice? Does legalized abortion allow everyone to make a free moral choice?
10. Albert Schweitzer once said, "If man loses reverence for any part of life, he will lose his reverence for all life." Is abortion a "violent solution" to an unwanted problem? What could happen to the values of a society which routinely destroys fetuses? Could abortion be but a step toward legalized euthanasia, as many believe? Could it lead to a "clip a flower in the garden" philosophy? Could the next step in such a trend be to get rid of the senile, infirm, retarded, etc.? Could the same ethical justification for taking the life of a fetus be applied to any other stage of life?
11. There are many nations in the world which currently have a higher abortion rate than the United States. In fact, some say abortion has become one of the world's largest "growth industries." Where might this trend end and what consequences might it produce on the way?
12. To what extent will the development of safe, efficient, and convenient chemical abortifacients enhance or aggravate the questions and problems just listed? To what extent would such an agent be used? Abused?

Control of Sexual Desire

1. Some psychologists believe that the sex drive is man's most basic and most powerful urge. Others disagree. However, *where* it is on the list is probably irrelevant. It should be obvious that the potential consequences which might result from the legal and/or illegal use of just negative sexual desire control agents alone can probably only be limited by the extent of one's imagination. When one adds the potential consequences of increasing the sex drive with positive sexual desire

agents, and promoting "instant intimacy" on all societal levels, a truly mind-boggling, almost incomprehensible scenario of the future begins to develop. To what degree would such agents be used and abused, legally and illegally? To what degree could such agents change and reshape the future?

2. A December 19, 1975 Vatican document titled "Declaration on Certain Questions Concerning Sexual Ethics" cited an "urgent need" to refute "serious errors," "aberrant modes of sexual behavior," and "the unbridled exultation of sex." The document was essentially a reaffirmation of previous, similar Vatican statements which have condemned adultery, premarital sex, and homosexuality. How would positive and negative sexual desire control agents affect such behavior?
3. According to some estimates, there may be as many as 500,000 full-time prostitutes in the United States today. Some estimates place the number of adult males who have visited a prostitute at least once at 80%. Would the use of positive and negative sexual desire control agents increase or decrease prostitution?
4. Should the use of positive desire control agents be declared illegal for certain segments of society such as the mentally-ill, the genetically-ill and deformed, the retarded, and convicted sex criminals? Should such segments of society be legally forced to use negative desire agents? Should the use of positive desire agents be declared illegal for those under 18?
5. LSD was originally intended to be a legal, clinical, research tool, but it "leaked" into society, and its subsequent illegal use eventually largely cancelled its intended use. Overall, could the development and legal and/or illegal use of sexual desire agents produce individual and societal physical, mental, and emotional problems far in excess of the problems which LSD and other drugs have produced?

Artificial Insemination

1. What are the legal ramifications of artificial insemination? To what extent does the use of AID change the meanings of such words as adultery, heir, illegitimacy, child support, and incest? Some states have passed laws guaranteeing the legitimacy of a test-tube child if its mother's husband gave his consent for the procedure, but most others have not. And, thus there have already been legal rulings of illegitimacy, with denial of custody rights, child support, etc. in many states.
2. In June of 1975, newspapers carried the account of a man in San Francisco who had just sued a commercial sperm bank for accidentally destroying his sperm specimen after he had a vasectomy. The

man asked for $5 million in damages for himself plus $500,000 for any other donor who had suffered the same loss.

3. Paternity is now defined as involving conception during the involved man's lifetime. What kind of legal or other problems will arise with the definition of paternity when conception is produced from the frozen sperm of a man who has died?
4. What kind of emotional and/or psychological problems are prompted in children who have been sired with sperm from anonymous and/or dead donors?
5. In 1974, two young people growing up in the same town cancelled their marriage after a consultation with their family doctor in which it was discovered that they were actually half-brother and half-sister.[2] They were both test-tube babies who had different mothers, but by sheer "coincidence" had the same anonymous sperm donor as a father. Could such bizarre "coincidences" increase in the future to the point where the genetic trouble that they produce could become significant?
6. Will society be able to resist the temptation to move artificial insemination from an anonymous to a selected donor basis? To use the sperm of famous people? What will be the individual and societal effects and consequences of choosing the general, and perhaps even specific, characteristics of your offspring? Should artificial insemination be used to practice negative and positive eugenics?
7. Should sperm be stored for the possibility of rejuvenating the human race with genetically undamaged children after a nuclear war? Whose sperm should be stored for possible future use?
8. In December of 1975, a group of about 30 persons demonstrated outside of Massachusetts General Hospital in Boston, calling for male supremacy and an elimination of sperm banks. The group, which called itself MS (Male Supremacy), has a red-maned white stallion as its symbol. It is apparently a whiplash movement, formed in opposition to women supremacists, who are feared to be able to control life and eliminate men if they were able to gain control of sperm banks!

Choice of Sex in Offspring

A slightly higher birth rate for boys is still being balanced by a higher mortality rate for males. However, choice of sex techniques were initiated clinically in 1976. It seems reasonable to predict that their use will increase greatly within the next decade.

1. Will history once again repeat itself? If given a choice, will society again choose to increase its male to female ratio? Would a trend toward more males also produce more learning problems, more need

for child guidance, more juvenile delinquency, more gangs, more illiteracy, fewer church-goers, more democratic voters, more suicides, more violence, more wars—overall, a more aggressive and more frontier-like society? Would more males also eventually produce a society wherein women are protected by taboos, and are treated like queen bees or ants? Could such a society also produce polyandry?

2. Could a tendency or trend toward more females than males produce trends largely the reverse of those just listed, and perhaps a more matriarchal society—a society which might also eventually practice polygamy?

Artificial Inovulation

1. What are the legal ramifications of artificial inovulation? How will the meanings of such words as heir, illegitimacy, and adultery change if the technique is widely used?
2. The American Medical Association has expressed fears that artificial inovulation techniques could lead to serious abnormalities in the fetus being grown and even to its sacrifice.[3] What new moral and ethical questions are involved in the abortion question if a fertilized egg or embryo can be implanted in a woman other than from whom the egg came? What criteria and attitudes should society adopt with regard to "weeding out" embryos with physical or biochemical abnormalities?
3. Will there be emotional and psychological effects in a woman who bears another woman's child? A child of another race?
4. Adding artificial inovulation to artificial insemination could produce a powerful technique for selecting general and even specific characteristics in offspring. Could such a technique also produce a trend toward using famous people as egg donors, practicing negative eugenics, and weeding "flaws" out of the human race? Could an initial desire to produce a "super race" turn into an irreversible compulsion to do it?

Artificial Placentas

1. What new moral and ethical questions are involved in the abortion question if an embryo can be transferred to an artificial placenta?
2. Studies with monkeys have indicated that separation of mother and child soon or immediately after delivery can significantly lessen maternal behavior, damage the child psychologically, and permanently alter what might be called the "mother-child bond." Other studies seem to indicate that the reverse is also true: increased contact after birth can change a mother's maternal behavior in positive ways.[4] What specific psychological effects result in a fetus from having spent nine months

in a mother's womb? Will those effects be changed with a nine month stay in a machine? What will be the societal effects of a number of people not only feeling that machines not only control their lives, but, to a degree, were their "mothers" as well?

Cloning (Parthenogenesis)

1. What will the legal ramifications of cloning be? What are the legal rights of a person's duplicate or set of duplicates? Will a nightmare of legal and illegal problems develop as an increasing number of duplicate people and sets of duplicate people exist in society? As a historical lineage of duplicate people is established?
2. What will be the military, political, economic, and scientific overtones produced through the possible cloning of large groups of identical people—people who are essentially immortal? For example, what would be the consequences should a certain government(s) establish a set of identical people in all of its key offices, a set of identical scientists working on the same weaponry problem, an army of identical, cloned soldiers, etc.?
3. What will be the cultural effects of cloning in society? What would happen, for example, to our natural, perhaps even instinctual competitiveness, if our most famous athletes, painters, musicians, and thinkers were cloned? Would two, three, four, or even five Kareem Abdul Jabbars on the same basketball team really be a step forward in sports, for example?
4. Considering the physical and mental "closeness" of twins, would cloning produce a more or less pluralistic society? Produce more or less patriotism and nationalism? Would it be easier or more difficult for a democratic nation to function and exercise governmental control of society, if that nation contained or was even made up of a certain number of blocs of people who had the same physical characteristics, temperament, character, personality, and perhaps even mentality? Might not a totalitarian state actually try to establish such clones in order to gain and maintain control of its peoples? How might revolutionary movements attempt to use cloning?
5. Individuals in the same cloned group will have certain physical advantages over non-cloned people, as in the case of transplants. Is such a situation fair? Will cloning increase a tendency toward a "have-have-not" society?
6. The cloning of humans will undoubtedly remain a rather complicated technological process and a rather expensive process. Should cloning only be available to the rich? How will the decisions be made as to who should be cloned? Who should decide?

7. The development of cloning techniques could result in a society requiring fewer males (or females) or even no males (or females) at all for reproduction. Theoretically, the earth could be populated by a self-reproducing society of women or men. What consequences could such a trend or attempted development produce?
8. Is humanness determined by the presence of a "soul"? If so, are "souls" transferrable? Do cloned people have "souls"? If not, are cloned people human? Can cloned people be "saved"?
9. Could a group of cloned people actually be described as being an unreal, unnatural, artificial assembly of immortal dolls all produced from the same mold—"xeroxed" human beings? A loss of a sense of identity would be expected in such a situation, and an undermining of individual expectations and aspirations. Would cloning thus produce a new type of complacent passivist simply waiting for the same, familiar ancestral scenario to unfold—a person for whom competition, fulfillment, wonder, and praise, have lost their meaning? Would cloning turn human beings into sheep? Sheep not all that concerned about possibly being led to the slaughter?
10. In cloning, "misfits" are not "junked" by evolution, but are actually duplicated. To what extent would cloning promote gene pool contamination? To what degree could cloning be historically dangerous?

General Questions and Problems Created by Reproduction Research

Our laws, literature, politics, economics, and, in fact, our whole social structure is based on the idea of two parents, who are in love, uniting sexually within a marriage to produce a child born of its mother's body. The possibility of completely removing childbearing and reproduction from sex and marriage raises the following three very disturbing, general questions.

1. Does the process of sex, as it has been genetically woven into man for three million years or more, also form the pattern for all the processes of life? To what extent are we changing that pattern today through the sexual revolution and reproduction research, by altering, amending and even changing the meanings of such words as male, female, love, and reproduction? Is it wise to reweave the pattern as extensively as we apparently are, in such a relatively short period of history? Overall, as we seem to increasingly reject "love" as a meaningful concept, place more and more and more emphasis on personal sexual pleasure, remove the idea of good and evil from sex, and classify monogamy as being monotonous, are we really turning sex into a more "natural" function, as many claim, or actually into a more unnatural function? Are we removing a potent set of neuroses in society, or simply generating a new set?

2. The "family," as we have known it traditionally and historically, is apparently in deep trouble. According to the 1970 U. S. Census, only 37% of the American people were living in a traditional or "nuclear family" consisting of working father, mother at home, and children. In fact, only 19% were living in a monogamous nuclear family in 1970. Those figures are undoubtedly even lower today. Most people today thus live in unmarried parent, single parent, remarried parent, dual working parent, foster parent, experimental, unconventional, or childless households—or they live alone.

   Some have argued that the deterioration and disappearance of marriage and the family are actually evolutionary trends, and/or necessary trends, and are thus natural and/or desirable phenomena. But are such beliefs historically accurate? To what extent has the family been the foundation or mainstay of civilization? Do societies without a family structure remain primitive? Do societies that begin with a family structure and then weaken or abandon it disintegrate, collapse, or turn totalitarian? Has any society survived unchanged or even survived at all after its family life deteriorated? Was the breakdown of the family unit actually one of the prime causes for the fall of the Roman Empire, as many historians contend? Is one of our basic, genetic, universal needs the need to be surrounded by a circle of at least several people that we can trust and lean on? The family usually fulfills that need with a father, mother, brothers and sisters, grandparents, and relatives. In the absence of the family, is it more difficult to establish one's circle in society at large? Is the family the best place, and often the only place, where one can find a sense of trust, discipline, and love that one can count on, and be cherished for who rather than what you are? Is the breakup of the family thus responsible for an increasingly serious epidemic of loneliness in the U. S.? To what extent is loneliness becoming a global problem? Was biologist Leon Kass right when he said several years ago, "The family is rapidly becoming the only institution in an increasingly impersonal world where each person is loved not for what he does or makes, but simply because he is. Can our humanity survive its destruction?"[5]

   To what extent is the family still alive, well, adjusting to, and surviving the tremendous pressures of our age? Will it be able to "hold its own"? How much more pressure can it take? Will it be able to survive the onslaught of new contraceptives, sexual desire agents, chemical abortifacients, choice of sex agents, test tube babies, artificial inovulation, artificial wombs, cloning, and "baby factories"? Or could reproduction research be the "straw that broke the camel's back"?

3. Anthropologist Margaret Mead recently said, "We have become a society of people who neglect our children, are afraid of our children, find children a surplus instead of the raison d'etre of living."[6] It is now beginning to appear that Dr. Mead may have actually understated the situation. Subjected to the pressures of increasing divorce; runaway parents; working mothers; poor parenting; alcoholism; drug abuse; materialism; physical, psychological, and sexual abuse; and various other societal factors, the parent-child bond seems to be weakening at an alarming rate. Recent evidence of the weakening is rather frightening. There may be as many as 120,000 children now in foster homes, with up to one million other children currently running away per year. There may also be as many as 250,000 "throwaway teenagers" per year—teenagers who have literally been ejected from their families.[7] Teenage drug abuse is increasing and there may now be as many as 500,000 juvenile alcoholics. Juvenile delinquency is increasing so fast that an estimated one in nine juveniles will be in juvenile court before they are 18. And suicide is now the second leading cause of death of those between the ages of 15 and 24.

   It would obviously be a serious mistake to completely put the blame for such statistics solely on parents. It does appear, however, that to some indefinable degree there is a subtle societal rejection of the young underway. Is that rejection being reinforced by a lowered birth rate and by a resultant increase in the average age of society? Will reproduction research also serve as a reinforcement? What will happen to our attitudes toward, and love for children when childbearing can be removed from sex and the meanings of such words as "mother," "father," "parents," "family," "brother," "sister," and even "children" become blurred? What will happen as we more and more lose the warmth, affection, and subtlety of intercourse and increasingly substitute a chemically driven and controlled liason? Could a situation actually eventually result such as that described in 1932 by Aldous Huxley in *Brave New World*—". . . leering obscenely at Lenina, and speaking in an improper whisper, 'Remember that in the Reservation, children are still born, yes, actually born, revolting as that may seem'"?

   What will be the psychological effects on a child when it knows it was born of geographically distant or even long-dead parents, and, when overall, the parent-child relationship becomes indistinct? Will the phrase "individual identity" lose much of its meaning? Will identity crises arise? Will the psychological problems of orphans and love-deprived children become more predominant? Is there a correlation between the childhood deprivation of love and identity and crime and a relentless drive for power? How many of those who have sought, obtained, and abused power through history have been what might be termed "loveless leaders" suffering from what might be termed the "Phaethon Complex"? (Phaethon was a mythological figure who doubted his divine origin and demanded that he be allowed to drive the sun's chariot for one day as proof. He nearly destroyed both himself and the earth in the process!)

The poor treatment of children has been commonplace through history. Swaddling, mutilation, disfiguration, and even infanticide were common well into the Middle Ages, and still exist even today in some primitive societies. Could reproduction research reinforce trends already in operation, to cause a historical swing of the pendulum back in that direction? Columnist Ann Landers conducted a poll in early 1976 in which she asked whether those who had children would do it again if they had it to do over. Over 70% of those who responded to the survey said no!

PHYSICAL MODIFICATION

Transplanted, Artificial, and Regenerated Body Parts

1. Is a "living" person a beating heart or a thinking brain? When does "death" occur? Can death even be defined if an artificial heart can be produced, powered by a nuclear battery, which will go on beating indefinitely, for all practical purposes?
2. Mechanical pumps, respirators, drugs, and tubes have allowed great extensions of life. Does a person have a right to live longer than nature would have allowed? Does one have a right to "death with dignity"? When should "the plug be pulled"? Who should decide?
3. What does the word "body" mean? Does a person have a right to sell his body parts either before or after he dies? Does one have a right to bury body parts that may be useful to or even sustain life in others? Does a person have a duty to donate his body parts after he dies? Before he dies?
4. There is currently a continual shortage of body parts. Who should get the spare body parts available or be able to use the techniques available? Who should decide?
5. Will transplantation and other means of renewing body parts lead to an intensification of a "have-have-not" society as more and more people desire to make use of such developments? For example, will the urban, rural, and third-world poor be "left out" again?
6. What are the implications of transplants between persons of different races?

7. What legal questions might arise if ovaries and testes can be transplanted? What psychological problems?
8. There have been extensive psychological problems thus far among the recipients of transplanted organs. With increased use of such techniques, what identity crises will develop? Will a person increasingly tend to lose the feeling of "uniqueness" and "identity"? Will one ask as never before, "Who am I" and "What am I"?
9. What does it mean to be your brother's keeper if you can, to a remarkable extent, control his life and death?

Genetic Engineering (Eugenics)

Some very difficult choices loom ahead for humanity, regardless of whether genetic engineering ever becomes a routinely used technique or not. For example, H. V. Aposhian has estimated that 4.5 times as many future life-years are lost as a result of birth defects as from heart disease, 9 times as many as from cancer, and over 10 times as many as from strokes. In fact, genetic disease may cause up to 25 percent of all hospitalization in the U. S.[8] In view of such statistics, should society continue the current medical practice of keeping everyone alive as long as possible? This policy not only results in increased medical problems and costs in the short-term, but allows many of those with defective genes to live long enough to pass them on to future generations, thus creating increased contamination of the genetic pool. An act of compassion to one generation can be an act of oppression to the next!

If society were to decide to attempt to decrease further contamination of the genetic pool, the question would immediately arise as to how it should be done. If society decided to continue to intervene to keep the hereditarily-ill alive, would it not then have to be willing to intervene again and force them to practice birth control or even undergo sterilization? Or should society decide to deny medical help to those with genetic defects? And what of more subtle questions? For example, what are the moral and ethical implications of building nuclear power plants today which will produce at least some genetic damage tomorrow?

1. Is the answer to the genetic contamination question to embark on a road of "preventative" or "corrective" genetic manipulation? If expertise and resources are limited to the practice of negative eugenics, which genetic illnesses should be attacked first? In some cases, harmful genes can also apparently serve beneficial roles. For example, the gene that can lead to death in sickle-cell anemia also apparently plays a role in protecting man from malaria.[9] How will that sort of situation be resolved? Should negative eugenics be voluntary or compulsory for those on whom it should be practiced? Should parents have the sole right to decide whether genetic manipulation should be performed on

themselves or their children? Or does society, the government, and/or the scientific community have a right to make that decision? What rights do unborn children and future generations have in such a decision? A small but increasing number of children are currently filing lawsuits against their parents because they have genetic disorders![10] Could we eventually reach a point where babies might have to pass a genetic acceptability test to be born? What would happen to those that didn't pass?

2. Current "recombinant DNA," "gene cloning," and "plasmid" experiments are a prerequisite toward the eventual development of negative eugenics. But such experiments are also now frequently producing bacteria and genetic material that do not exist in nature. In June of 1976, the National Institute of Health published guidelines designed to minimize the risk of spreading such material through the environment, and especially colonizing the human intestinal tract (many of the experiments involve *E. coli*) or any other part of the body. The containment is supposedly to be accomplished through physical precautions within the laboratory and by disabling the genetic material produced in such ways that it can only propagate under specialized laboratory conditions. Most of the scientists involved in such experiments claim the guidelines assure that there is only a remote or minimal chance that society could be endangered by an accidental release of such material. However, they will not say that the odds are zero. And it should be pointed out that as more laboratories perform such experiments, the odds favoring accidental release increase accordingly. Should such a release actually occur, the results could conceivably prove calamitous. Depending upon its genetic nature, such material could conceivably introduce humans to new diseases and lethal toxins for which we have no natural defense systems and thus would be helpless. Or, such material could interfere with our immunological system, lower our resistance to drugs, create enzyme or hormonal imbalances, or even create contagious cancer.[11]

The atomic bomb, pollution, nerve gas, and the destruction of the ozone layer are all threats to human existence. But they can all be limited or controlled—in theory if not in practice. Has recombinant DNA opened a new door in this respect? Was Dr. Liebe F. Cavalieri, a Cornell University biochemist, right when he recently said, "The threat of a new life form is more compelling, for once released, it cannot be controlled, and its effects cannot be reversed. A new disease may simply have to run its course, attacking millions in its path. . . . This research is the greatest threat ever to our human experience."[12]

3. To what degree would the *intentional* synthesis and subsequent release of new genetic material not found in nature increase the "minimal" dangers just outlined? Our current world of decreasing inhibitions, morality, and ethics, but increasing mental, emotional, and psychological imbalance, seems to be spawning two alarming new trends. First of all, an increasing number of individuals and groups of individuals are openly and anonymously engaging in terrorism, societal blackmail, and even death. To what extent might genetic research reinforce that trend and perhaps even usher in a whole new era of such activities? Perhaps that era is already here. Some still believe that the mysterious "legionnaires' disease" in Philadelphia in 1976 may have been caused by an individual or group of individuals who released a rather sophisticated chemical or biological agent at the convention! Secondly, in an age in which most nations can't afford the preparations needed to fight a massive, overt war, and such war can only result in mutual self-destruction anyway, it should be obvious that the ability to wage secret, covert warfare might become increasingly attractive to all. In 1974 the United States, Russia, and other nations agreed to destroy their stocks of biological weapons. However, in 1975, a congressional investigation disclosed that the CIA had defied a presidential order to that effect. Many believe that Russia never discontinued such research. In March of 1977, the United States Army publicly revealed that it had conducted 48 public tests of biological agents in the 1950's and 1960's.[13] The agents were believed to be harmless at the time, but some were later identified as being potentially dangerous pathogens.

Could genetic research open, reopen, or reinforce biological warfare research around the world? Dr. Cavalieri recently said, "It is possible, intentionally or unintentionally, to construct highly dangerous agents of other types, worse than anything yet envisioned in biological warfare."[14] Gordon Rattray Taylor underscored the possible desirability of such research in 1968, when he said,

> In current thinking, the best way to wage war is to wage it without your enemy even being aware it is happening. . . . If viruses could be used to carry new genetic material into cells, perhaps one could tamper with the genes of another nation without their ever realizing the fact. History would simply record, as it has so often in the past, that such and such a nation rose to power while certain other countries entered a decline. . . . Perhaps there are nations consciously waging this kind of warfare now.[15]

Many believe it is only a matter of time before plasmid experiments are performable on a high school level. If only one person in a million were to attempt to synthesize and/or use genetic material for evil purposes, there would be 4,000 people in the world actively engaged in such activities!

4. Some believe negative eugenics may someday be used to change biological rhythms and "clocks." What negative societal consequences might result if genetics can be used to tamper with or change cell proliferation and division, the sleep cycle, menstrual cycles, the instincts of animals and insects, etc.?
5. Could the development and practice of negative eugenics eventually also produce an era of positive eugenics? If man can someday redesign as well as "correct" himself, what should the new design be? Who should the architect be? If man becomes "superman," will he become more or less human? What specific problems could arise? For example, animal and human sporting events play a major role in society. Would positive eugenics put an end to such events or simply raise them to undreamed of levels?
6. In what ways might both negative and positive eugenics be used for racist purposes? For example, might individuals or even large groups of people attempt to change their skin color and/or physical characteristics? Could such techniques be used in another historical attempt to build a "super race"?

Artificial and Synthetic Plants and Animals; Man-Animal, Man-Plant, and Plant-Animal Chimeras

Research in these areas is essentially in its infancy. It is rather difficult to predict where current research will lead and within what time scale. It is also thus probably premature to attempt to predict possible specific negative consequences for this work. However, assuming that considerable progress will be made in one or more of the areas, the following general questions come to mind.

1. What would be the moral, legal, emotional, and psychological identity of a chimera which was partly human? How man-like would something have to be to be classified as being human? Would the decision be based on appearance, chromosome count, or both?
2. What evolutionary effects could such advances produce? What could happen to the "balance of nature" and the competition, complementary, and balance between life forms currently found in nature?
3. What would such advances do to man's reverence for life and his belief in a natural order and mutuality of things?

MENTAL MODIFICATION

The Electrical Control of the Brain

1. Is a person's identity completely determined by the brain? Where is

the "seat" of one's humanness located? What guidelines should be established for research involving the brain?

2. Is it "right" to "wire" people for pain-relief, happiness, and pleasure?
3. Could research on stimulating the brain electrically lead not only to behavior control but also to *absolute* torture techniques? Could such research lead to a loss of democracy and the intensification of a "have-have-not" society?

The Chemical Control of Behavior, Memory, and Intelligence

1. What does it mean to be "human"? What is the "human console"? How do we strike a responsible balance between artificial and natural mental processes?
2. Is the deliberate, effective, and possibly irreversible chemical and/or electrical shaping of human intelligence, memory, behavior, and personality morally and ethically different from the piecemeal, accidental, and often ineffective pressures of schools, family, and so on? If so, how?
3. In altering intelligence, memory, behavior, and personality, whose set of values, and whose criteria should we use? Who should decide what an individual should become? What society should become?
4. What psychological consequences will result from intelligence, memory, behavior, and personality modification? How will previous interpersonal relationships be affected? What legal problems will arise (e.g., responsibility for previous crimes)?
5. A human being is a blend of intelligence or thinking and emotions. Could the balance between thinking and feeling be upset by mind research? Will mind modification breed more or less human or inhuman people? We frequently feel "awkward" around intoxicated or insane people. How disturbing will it be not to know to what extent a man is genuinely himself?
6. *Clockwork Orange* was a very interesting novel and motion picture. In the story, a delinquent who obtains sexual excitement from violence, destructiveness, and rape is conditioned to feel sick at the sight of violence. He actually becomes meek and fawning, "artificially." Is there any value in virtuous behavior which is not based on moral effort and moral choice?
7. The use of drugs in the U. S. and the world has already become a scourge, although many or even most people now consider such usage to be "normal." In the early 1960's, thalidomide was considered to be a "harmless" tranquilizer and sleeping pill, until it produced a lost generation of about 10,000 misshapen children. According to some estimates, about one fourth of all prescriptions given each year in the U. S. are for drugs that either relax or stimulate the body, as many

as 30% of all women and 10% of all men now use some kind of tranquilizer, about 80 million Americans use alcohol, etc. To what extent would society legally and illegally abuse mental modification and behavior modification techniques?

8. Are chemical and/or electrical mental shaping techniques so inherently coercive that they threaten individual freedom? Under what conditions would it be acceptable to alter the consciousness of an individual? What decisions of this type should the state have an influence on? For example, should parents or an outside agency be primarily responsible for the values of the young? Should society be free to alter the basic values of criminals and others with "anti-social" traits, impose new values on a ghetto child in an attempt to get him "up and out," etc.? Should society provide a "crutch" to some people by making their decisions for them? Could this lead to brainwashing and mind control? If society or the state can control the consciousness of individuals, who will control the controller?
9. Could research on hemispheric dominance be combined with research on the age-old question of whether man actually has free will to produce a chemical and/or electrical method of interfering with or even controlling that free will?[16]
10. If presidents, congressmen, etc., were forced by various means to use "peace pills," what would prevent their use on an unknowing general populace? What would prevent their abuse by secret police and intelligence agencies?
11. If memory transfer is accomplished through DNA and/or RNA injections or transfer, who will decide who the donors should be? What will be the effects on memory-associated skills? Could such skills be transferred? What will be the individual and societal effects of memory erasement or inducement? Is the falsification of history possible, for example?
12. To what degree will genetic engineering, physical and mental modification, and reproduction research produce combined or even synergistic societal effects? For example, in *Brave New World,* Aldous Huxley imagined a society free of war, poverty, class conflict, disease, brutality, anxiety, frustration, and inequality. But the price was the disappearance of thought, love, friendship, science, poetry, judgment, curiosity, sentiment, and beauty!

Disembodied Brains, Head Transplants, Brain Transplants

1. What criteria should be used in deciding whose brains will be disembodied and whose brains and heads will be transplanted? Who should decide what those criteria should be?

2. What would the legal ramifications be of the use of such techniques?
3. Is the concept of a human being having a "soul" meaningful or meaningless? If a person does have a soul, where is it—in the brain, in the body as a whole, or in both? Does a disembodied brain have a soul? If the brain or head of one man were transferred to the body of another, would the resultant structure have one soul or two? Who is the survivor in a head or brain transplant? Will the phrase "resurrection of the body" have any meaning if head and brain transplants and disembodied brains become possible?
4. A disembodied brain may need the blood supply of a healthy person to survive. Would it be "right" to use a healthy human to keep another's brain alive? What psychological effects would result if the disembodied brain was simply hooked up to a machine?
5. What psychological problems might result with the use of such techniques, for society in general as well as for the individuals involved?
6. At what point, if any, do things such as a "brain in a bottle" become a travesty or mockery of nature and the evolutionary process?

Man-Computer and Man-Machine Chimeras

1. There are no computers today which can think or reproduce themselves, as we now define those terms. Many believe such computers will never be developed. But others are not so sure. As Marvin L. Minsky of M.I.T. stated in 1966:

> Once we have devised programs with a genuine capacity for self-improvement a rapid evolutionary process will begin. As the machine improves both itself and its model of itself, we shall begin to see all the phenomena associated with the terms "consciousness," "intuition," and "intelligence" itself. It is hard to say how close we are to this threshold, but once it is crossed the world will not be the same.
>
> It is reasonable, I suppose, to be unconvinced by our examples and to be skeptical about whether machines will ever be intelligent. It is unreasonable, however, to think machines could become *nearly* as intelligent as we are and then stop, or to suppose we will always be able to compete with them in wit and wisdom. Whether or not we could retain some sort of control of the machines, assuming that we would want to, the nature of our activities and aspirations would be changed utterly by the presence on earth of intellectually superior beings.[17]

Is HAL, the thinking and talking, but mutinous computer in *2001—A Space Odyssey,* actually a possibility for the future? If thinking, reproducing computers are built, what will they need man for? Could MAN someday stand for "meaningless," "archaic," "non-entity"?

2. Is it possible that computers might be produced one day with their own personalities, or will human and computer "thinking" processes always remain fundamentally different? Could a mini-computer implanted in the brain develop independent notions of its own, "seize control," and turn its human host into a robot? Or would it always retain the same relationship to its owner as, for example, one's eyeglasses? To what degree could such a subservient computer be contacted, influenced, or even programmed by other persons than its owner, for evil intentions? Adam V. Reed has stated: "It's foreseeable that some dictator, with access to the internal codes of memory, could find out what people are thinking. We should begin thinking of these problems now, before they arrive."[18]
3. It should be obvious that even if computers, machines, and man-computer and man-machine chimeras remain subservient to man, the psychological, economic, political, and military ramifications which they could produce can only be limited by one's imagination.
4. What could such chimeras do to our reverence for life and our sense of a balanced "natural order"?

Mental Modification "Intangibles"

1. Stanford University Nobel Laureate William Shockley believes there is an "illusion of flat human quality" and that there are inherited intellectual differences between races that have a basis in genetic differences. In other words, Shockley (and others) believes that all men are not created equal, and that this can be proven genetically. He has warned of a possible "dysgenic threat"—regressive evolution caused by the disproportionate reproductive rates of people with lower intelligence. In the early 1970's, the Stanford Advisory Committee and the National Academy of Sciences refused to sanction further study into the matter. Were those decisions consistent with a spirit of impartial scientific inquiry, academic freedom, and the search for truth? If the work is done someday and it upholds Shockley's contentions, what societal and global consequences might result?
2. Is it really wise to create increasing physical and mental uniformity in society by simultaneously eliminating the weak and unfit and engineering new traits into the survivors? To what extent have the weak and the unfit actually molded history and developed the various attributes of society? Have not the weak and unfit not only usually survived but often triumphed over the strong? Would increasing uniformity and decreasing diversity in society actually rob humanity of much of its strength?

3. Many people have claimed that LSD will produce mystical and/or religious experiences, in which one can encounter "God."[19] Suppose other chemical agents were found which could heighten or enhance such experiences. Would a realization that the God concept is apparently nothing more than brain chemistry strengthen or weaken that concept? What questions would such a realization prompt for the theologian? For the atheistic or materialistic scientist? What would be the societal consequences of the wide-spread use and/or abuse of such agents?

THE PROLONGMENT OF LIFE

Control of Disease

1. A big part of man's current success in controlling disease and extending life-spans can be attributed to the increasing use of antibiotics. But that increasing usage is also apparently producing a vicious cycle in which the antibiotics produce stronger germs, necessitating the development of stronger antibiotics.[20] As Dr. Stanley Falkow, professor of microbiology at the University of Washington Medical School has said, "Man develops a better mousetrap, and nature develops a better mouse."[21]

   Unfortunately, there also seems to be growing evidence that the use of antibiotics as growth hormones in cattle, swine, and poultry may also be promoting the same vicious cycle.[22] It now appears that a multi-million dollar industry which has helped feed the world's population since the 1950's has also helped human bacteria build resistance to antibiotics. Overall, is man producing an era of drug-resistant "supergerms" through his indiscriminant, imprudent, and prolonged use and abuse of antibiotics? Could man be headed for drug-resistant epidemics of worldwide proportions? Perhaps of even once-conquered infectious diseases? Was Dr. Richard P. Novick of the Public Health Research Institute in New York City perhaps right when he recently said, "It's a question of who runs out of ammunition first—us or the bugs." (See footnote 21.)

Freezing Techniques

1. If people can ever be frozen for extended periods of time and then revived, what will happen to our legal system and the meaning of words such as "estate," "life insurance," "social security," "pension," "back taxes," "trust funds" and "property"?
2. If specified in a person's will, would failure to freeze or revive someone be murder? Could cryogenics interfere with autopsies and complicate the detection of murder?

3. Could cryogenics interfere with a person's right to be buried or cremated as he wished?
4. If the term "soul" has any meaning, where was a person's soul during freezing?
5. What psychological problems and cultural shock would be suffered by persons frozen and then revived after long periods of time?

General Questions and Problems Created by Prolongment of Life Research

1. How will a person's ability to live longer affect the population problem? Would extending the life-span necessitate a limitation or even a cessation of procreation in order to maintain a steady-state population? What balance should be struck between the right to live longer, the right to reproduce, and the rights of life as yet unborn? If reproduction is limited, who should have the right to bear children? Where is the balance point between one person living longer versus several people living less?
2. Will prolonging life also prolong its boredom, apathy, and ennui? How will the chief source of old-age boredom be relieved, a feeling of having experienced most or all of what life has to offer? What will prolonging life do to interpersonal relationships? Would extended life-spans lead to marathon marriages or simply make married life increasingly unbearable? Is it the prospect of death which actually makes life bearable and tolerable to at least some degree?
3. As life-spans increased how would one's survival and preservation instinct and his "will to live" be altered?
4. Does a person have a right at a certain point in his life to a "death with dignity"? Could the extension of life-spans produce a society where the right to die might have to be defended and even fought for? Could a society result where one might need a license to die? Could people in such a society who choose to die be considered to be guilty of suicide? Who "owns" your body and your life? Yourself, society, doctors, God?
5. The aging of the body and the aging of the brain are apparently different physiological problems. What problems could arise in aging research because of that fact? For society?
6. What societal problems will be generated by an extension of the life-span? Will the extension be the same for men and women, or proportionately longer for women? Could that produce an increasing number of women in society and increasing trends toward matriarchy and polygamy? What will happen as the elderly constitute a larger and larger proportion of society? Only about 1 in every 25 people in the United States was over 65 in 1900, but today more than 1 in every 10

are. According to the U. S. Census Bureau, the number of people over 65 has doubled since 1960. Will increasing life-spans and a growing number of elderly dampen the country's "frontier spirit," prompt a trend toward later retirement and thus fewer opportunities for the young, put increasing pressure on young people to support social security, etc.? Could such problems promote increasing support for the following stereotype of the elderly: "Old people are grumpy, crotchety, doddering, decrepit, depressed, depressing, over the hill, all washed up, and unable or unwilling to learn new skills. They've seen better days. They're set in their ways. They're sexless, tired, irregular, paranoid, and neurotic. And, of course, they're all alike."[23] Could such feelings of rejection gradually intensify into an actual revolt against the old? Could the revolt produce a trend toward euthanasia? Obligatory euthanasia? Can a society in which the old "overburden" the young survive?

Or could the reverse occur? Senior citizens are 10 percent of the population in the United States today, but 20 percent of the poor! In fact, they usually have more problems of almost every kind than anyone else—health, housing, transportation, loneliness, boredom, responsibility, dignity, etc. As their numbers swell, will the elderly become more vocal and demanding and seek redress for the wrongs done them, perhaps even to the point of revolt?

7. Control of the aging process could conceivably put immortality within our grasp. Would immortality be a blessing or a curse? The punishment bestowed on the betrayer of Christ was immortality!

THE CREATION OF LIFE

1. Creation of life research is currently directed toward producing life as we know it. But, once successful in this area, the scope of such research will undoubtedly turn toward producing new life forms. What kind of "accidents" and/or consequences might result as man begins to assemble amino acids in ways never before seen on earth?
2. Will the interchangeability of natural "living" and synthesized "living" matter erode the feeling that "living" matter is different from "non-living" matter? What will be the consequences of such an erosion?
3. As life becomes more "mechanical" and "manufacturable" will it also become "cheaper"? Will we lose our reverence or respect for life and the natural order of things? Will life lose its mystical significance? Should we allow the body to become only a piece of property—like an automobile or TV set? What societal consequences will result from all of this?

4. Research on the creation of life is increasingly indicating the probability that the same chemistry exists everywhere in the universe and that life also exists elsewhere in the universe. What societal ramifications do such ideas imply?
5. What are the dangers that research into the creation of life could result in the creation and release of an artificial virus which could loose a plague on mankind for which he has no bodily defenses? Might such research be used to wage secret warfare?

IN SUMMARY

Alfred North Whitehead once said, "It is the first step in sociological wisdom to recognize that the major advances in civilization are processes which all but wreck the societies in which they occur." The biological revolution has already produced one "major advance" in civilization—the "pill." It promises to produce a minimum of at least several more. To what extent could such advances "wreck" society? It could be argued, of course, that not all of the negative implications and consequences outlined in this chapter will develop into societal problems, and that some are actually quite remote or far-fetched because the developments that might produce them will never be brought to fruition. But it could also be argued that if only a fraction of those consequences which have been listed do so develop, they could cause a staggering amount of societal complexity, confusion, and flux. Societal problems could conceivably develop which could make those produced by the pill and abortion look like a Sunday school picnic by comparison! And it might also again be pointed out that the list just presented is incomplete.

Taking an overall look at the negative consequences which the biological revolution could conceivably produce reminds one of the following fable, first related long ago.

> To the amazement of their rude companion, three great magicians in turn caused a pile of bones to become a skeleton, then fleshed out a bony structure, and finally prepared to endow it with a new life. Shaken, the simpleton cautioned, "Don't you realize that these are the remains of a tiger?"
>
> Caught in the frenzy of artful invention, the magicians ignored the warning. From his tree, the simpleton watched the splendid moment of rebirth and—ultimately—the consumption of the three wise men.
>
> —Sanskrit *Panchatantra, A Fable*

CHAPTER 4

BIO-BABEL

> I do not pose as a preacher, but let me tell you that if there is a God, He will not let us advance much farther materially until we catch up spiritually. A fundamental law of nature is that all force must be kept in balance. When any force goes off on a tangent there is a smash.[1] —Thomas A. Edison

The statement was made earlier that "the biological revolution carries great potential danger for society, and perhaps even the threat of societal disintegration and collapse, through the additional, and yet largely unpredictable and unexpected complexity, confusion, change, flux, tension, and future shock that it could add to that which is already foreseeable and predictable." It was also contended earlier that "such a prospect cannot be simply dismissed" and "is at least debatable." After seeing at least some of the potential implications and possible societal consequences of the biological revolution outlined in the last two chapters, the reader should now be at least somewhat receptive to those contentions, no matter how optimistic about the future he or she may be.

But there is yet another chapter to be added to our unfolding story, for the more obvious, *primary,* societal problems outlined previously will also produce some broad, interrelated, overriding, *secondary* questions and problems—overeffects with a distinct moral, ethical, philosophical, and religious flavor. The secondary questions and problems will thus be even more nebulous, perplexing, and difficult to "get a handle on" than those described previously, because they focus more directly on questions of conscience, obligation, virtue, values, should or shouldn't, right and wrong, etc. In fact, the secondary questions actually become incredibly complex when one also adds the definitions and redefinitions that will be needed prerequisites for intelligently carrying out the biological revolution, for man has argued about many of those definitions since time immemorial.

The reader is again advised that the following two lists of general and specific secondary questions, problems, and needed definitions and redefinitions are not intended to be all-inclusive, but include only those questions and problems foreseeable by the author at this time. And it should also be emphasized that, because of the almost incredible complexity of the questions and problems posed, there are no generally accepted answers available to them as yet. Whether or not there ever will be is an open question at this time!

GENERAL OVEREFFECTS

1. WHAT IS A MAN? That question has come echoing unchanged down the halls of time over and over again, always arriving exactly as it was first asked eons ago. But when an answer has occasionally followed, it has always arrived garbled—a heterogeneous, incongruent, and largely undecipherable mixture of thought, of which one can usually understand only a word or phrase here or there. For example, man has been described as "a reasoning animal" (Seneca), "a prisoner" (Plato), "a noble animal" (Sir Thomas Browne), "a mere insect" (Francis Church), "a beast" (Thomas Percy), "born free" (Rousseau), "master of his fate" (Tennyson), "certainly stark mad" (Montaigne), "a naked ape" (Morris), "a territorial animal" (Ardrey), "a beast or an angel" (Dubos). After pointing out some of the above definitions, Floyd D. Matson, in his 1976 book *The Idea of Man,* goes on to say, "The definition of man—or, more exactly, the *identity* of man—is an open question, a live issue, up for grabs. It has never been settled or agreed upon; possibly it never can be."[2] But he does also add that many of the answers that have been offered fall into three classes of thought. He claims there is "the creature model: man as *beast*"; "the robot model: man as *machine*"; and "the creator model: man as a *free agent.*"

Is man only an animal, a slave to his unconscious, instinctive emotions, desires, and fears, as Darwin, Freud, Lorenz, Ardrey, and others have argued? Is he only a machine, forced into reflexive responses, incapable and even undesirous of taking responsibility for himself, as B. F. Skinner, Lionel Tiger, and others have contended? Is he only a weak, passive, limited, irrational, predictable creature, unable to exercise free will and incapable of handling the responsibilities of freedom? Or is he an active, resourceful, creative creature —a "free agent" as Matson and others have argued? The Austrian naturalist, Konrad Lorenz, gave perhaps one of the more interesting answers that has been given to such questions when he once said, "Man appears to be the missing link between anthropoid apes and human beings." What is man? *Man still does not have a comprehensive conception of what he is!* The answer is still "up for grabs."

2. WHAT IS MAN'S RELATIONSHIP WITH GOD? The question obviously has no meaning for those who profess not to believe in God. But for those who do, it can be a more thought-provoking question than it appears at first glance. About 2,500 years ago, the psalmist asked, "What is man that Thou art mindful of him?" The answers offered since have not been completely satisfactory! To what extent, for example, was man created "in the image of God" and a "creature of God"? What is our understanding of the word "God"? Is God an intelligent force that acts outside of nature or a force that operates from within? *Is* God nature? Is God simply the life-stuff of the universe—energy? Was the creation a static, Genesis-type of creation, or is it an ongoing evolutionary process? Was there an original "contract" between God and man with regard to the creation, or is it an open, ongoing relationship? In either case, what are the "rules of the game"? Has God revealed those rules? How? Where? What do we understand to be the wishes of God for his creation? Did God intend man to be "distinctively human"? If so, what does it mean to be "human"? What is the basis from which a human being operates? Did God endow man with a "mind" or does man simply have a more evolved brain than any other creature in nature? What do the words "body," "brain," "mind," and "consciousness" mean? Is a man more than the sum of his body parts, his chemistry, and his genetic determinants? Are the words "spirit" and "soul" meaningful or meaningless in this respect? Is man a duality of body and spirit or soul? If so, at what point in the historical development of humanity was a spirit or soul infused? At what point in the development of an individual human being is a spirit or a soul infused? What do the words "good," "evil," "sin," "original sin," "predestination," "free-will," "freedom," and "responsibility" mean in the context of a soul or spirit? How are they related to the God concept? What do they mean if there is no God? For example, if man was not born in original sin, how did he become the destructive creature we know him to be? Is his aggressiveness inherited, learned behavior, or both? The British statesman Disraeli, when once asked whether man was an ape or an angel, replied, "I am on the side of the angels." Rousseau and others have maintained that man was essentially good until he experienced civilization. However, the fashionable view today is that there is a fundamental bestiality in man which civilization has simply given a wide range of expression. Is man intrinsically good or evil? Why?

Is there a life-death continuum? What do the words "life" and "death" mean? Are the words "life after death," "resurrection," "resurrection of the body," "salvation," "Kingdom of Heaven," and "immortality" meaningful or meaningless? Should "immortality" mean an indefinite stay on earth or something reserved for the "life hereafter"?

Man is reluctant to unhesitatingly affirm the existence of God, has no comprehensive concept of God, and is unsure of his relationship with God!

3. WHAT IS MAN'S PLACE IN AND RELATIONSHIP TO THE REST OF NATURE? Is man really "distinctively different"? That age-old contention is being seriously and outrightly challenged by some modern scientists and inadvertantly and unintentionally undermined by others. As Francis Ensley said over a decade ago,

> The outcome of the scientific approach is to depreciate man. Astronomy proclaims his microcosmic size. Biology claims that he had animals, if not for parents, at least for first cousins, in the long evolutionary series. Chemistry affirms that he is a compound of hydrogen, oxygen, carbon, and other elements, of the same essential stuff as sticks and stones. Psychology teaches the equivalence of the species: that every basic feature of human nature can be studied without essential loss in the rat and dog. If we add to the theoretical degradation of science the fact that it has supplied the weapons whereby the human race can be liquidated, the indignity is complete.[3]

Or is it complete? Since Ensley spoke those words, biochemists and molecular biologists have mounted a full-fledged assault directed toward understanding and changing life, and have increasingly succeeded in blurring the distinction between the living and the non-living!

Is man a creature above and apart from nature, above but within nature, or simply a part of nature? Is man the "superior animal," the hub or axle in the wheel of nature, around which all the other "spokes" revolve? If so, what are non-human creatures for? Or, is man simply a spoke himself—although admittedly possibly a larger spoke? Is science taking us closer to, or further away from the answers to such questions? If there is a God, and man is a creature of God and distinctively human, is it valid to increasingly look upon man as simply being a collection of atoms and molecules, whose behavior is essentially explainable by the laws of chemistry and physics? If there is a balance point between scientized and humanized man, where is it? *Man does not have a comprehensive conception of his place in nature!*

4. WHAT IS MAN'S RELATIONSHIP TO THE EVOLUTIONARY PROCESS? If evolution is the primary process through which nature operates, is there a "blueprint" for evolution or is it simply a blind, random process? Is God the directing force behind evolution, does God work through evolution, or *is* God evolution? Or, is evolution simply a unique process unique to the earth, the end result of a "cosmic accident" in which life was formed by accident only here on the earth? Most people would undoubtedly agree that man is evolution's supreme product. But how much agreement is there as to *why* that's apparently true? Why is man apparently the only species aware that it evolves? Was Pierre de Chardin right when he once said that "Man is evolution grown conscious of itself"? Are man and human nature still evolv-

ing? Is this all by plan or by accident? The theory of evolution is still only a theory and evolution remains a largely indefinable theory at best. *And thus it is that man still does not have a comprehensive conception of his relationship to the evolutionary process!*

5. WHAT IS MAN FOR? WHY IS MAN HERE? Are those questions to be answered only in terms of short-term expediency and survival, or is the long-range answer to be more than that? Is man's role on the earth to be merely that of a sharecropper or caretaker, or is he either destined or predestined to be more than that? Does the fact that man is "distinctively human," "distinctively different," the "superior animal," "evolution's supreme product," and/or a "creature of God" indicate that he has a right and perhaps even a responsibility to be a partner, "master planner," or even "boss" in nature and the evolutionary process, making of it what he will? Are his powers in this respect destined or predestined to be limited or unlimited? Which men should make those decisions? Does man also have a right or even responsibility to direct or alter his own evolution and destiny? To a limited or unlimited extent? Which of his human values and characteristics should be changed? Which retained? Who should decide?

Such questions remain extremely complex, even if God is removed from them as a consideration. For example, suppose man did decide to proceed with becoming a partner, "master planner," or even "boss" in nature and the evolutionary process. Does he really know enough about evolution to intelligently pace himself? Biologist James F. Danielli made the following, rather astounding predictions about that possible pace in 1972.

> The age of synthesis is in its infancy, but is clearly discernible. In the last decade (1960–70), we have seen the first syntheses of a protein, a gene, a virus, a cell, and of allophenic mice. Nothing with such dramatic implications has ever been seen in biology before. Previously, plant and animal breeders have been able to create what are virtually new species, and have been able to do so at a rate which is of the order of 10^4 [10,000] times that of average evolutionary processes. A further increase in rate is now on the horizon. We need a few additional "firsts" before this will occur: (1) to be able to synthesize a chromosome from genes and other appropriate macromolecules; (2) to be able to insert a chromosome into a cell; or, alternatively to (1) and (2), to be able (3) to insert genes into a cell in some other way; (4) we must also learn how to bring the set of genes, which is introduced into a cell, within the domain of cellular control mechanisms, so that they do not run wild in the cell. None of these problems appear to be of exceptional difficulty. When these techniques are available, the possible rate of formation of new species will again be accelerated by a factor of 10^4 to 10^5 [10,000 to 100,000]. . . . It will be possible to carry out the equivalent of 10^8 to 10^9 [100 million to 1 billion] years of evolution in one year. It will be surprising if we do not reach this point within 20 to 30 years, and we may well be there in 10 years.[4]

To what extent would such a pace be "unnatural" and violate either a "blueprint" behind evolution and/or the natural, leisurely way in which it has traditionally unfolded? Should we actually attempt to "take the reins" from nature's hands in this respect? Montaigne once said we should "permit Nature to take her own way," for "she better understands her own affairs than we." To what extent was he right? Could it be unpredictably dangerous to remove nature's prerogative to set her own pace? To set her own course? Nature's course is variability. Man's course would undoubtedly be greater uniformity.

> Evolution is founded in variety and creates diversity; and of all animals, man is most creative because he carries and expresses the largest store of variety. Every attempt to make us uniform, biologically, emotionally, or intellectually, is a betrayal of the evolutionary thrust that has made man its apex.[5]
>
> —Jacob Bronowski

> It is foolhardy to eliminate genetic variability. That is our evolutionary bankroll, and we dare not squander it. Species that ran out of variability ran out of life.[6]
>
> —Gerald McClearn, geneticist

> In the long history of evolution, 100 million species of plants and animals have inhabited the earth. Of these, 98% are now extinct, unable to survive the challenges of a changing environment. Man himself may face such a life-and-death test. Unlike his predecessors on the evolutionary ladder, he has the capability to meet it—and to fail even more grandiosely than did creatures with lesser brains and imaginations.[7]
>
> —Quoted in *Time* magazine

> The chess-board is the world, the pieces are the phenomena of the universe, the rules of the game are what we call the laws of Nature. The player on the other side is hidden from us. We know that his play is always fair, just, and patient. But also we know, to our cost, that he never overlooks a mistake, or makes the smallest allowance for ignorance.[8]
>
> —Thomas Huxley

If one does attempt to consider God in the question of whether or not to become a partner, "master planner," or even "boss" in the evolutionary process, the decision obviously becomes much more complex and uncertain. For example, man is still uneasy about the lesson of Eden. "Eat of the forbidden fruit," God warned, "and you shall surely die." "Eat," promised the serpent, "and you shall become like God." Was that a one-time only warning, establishing a pattern wherein man was to be punished by God each time thereafter that he attempted to "become like God"? If not, what *is* the lesson of the Tower of Babel story? Did Eden establish man's free will to thereafter eat from the tree of knowledge of good and evil, with nature to punish him only if he "overeats"? Is it really part of the human task to improve and fulfill the intentions of God for the earth? To actually share the Godhead? Or, would such efforts actually be "tinkering" with the natural order and tampering with the divine plan? At what point, if any, do things like a "brain in a bottle" become a mockery of God, nature, and the evolutionary process? Wernher von Braun once said that space exploration must have God's blessings because he hasn't put any obstacles in its way. But would placing obstacles in man's way be incongruent with giving him his free will? If God doesn't place any obstacles in the way of the biological revolution, does that really mean that it has his approval? Do the questions that we are asking actually require answers that involve more than simply right or wrong and should or should not decisions? Are both of these positions merely extremes? If so, where is the balance point between man's efforts to fit himself into the evolutionary process, his attempts to control or even exploit it, and his sense of responsibility to limit his actions to a level acceptable to both nature and man?

Even if the human being is nature's most unique creature and is possibly a "creature of God," does that really give him a right to experiment on other, lesser creatures? Does he really have a right to subject 60 to 100 million animals per year to experimentation, often agonizing pain, and even death? To what extent should one human being be permitted to endanger another in the name of scientific and/or medical progress? Part of the Hippocratic Oath states, "I will follow that system of regimen which, according to my ability and judgment, I consider for the benefit of my patients, and abstain from whatever harm or injustice."

The Nuremburg trials reaffirmed the legal principle that no human being can be used in an experimental situation without his or her consent. Perhaps this is why Americans have been shocked in recent years to learn of the extent to which such experimentation has been taking place in our laboratories, prisons, hospitals, etc. In 1972, for example, a 40-year federal study came to light which indicated that at least 28 black Alabama syphylis victims died, and many others may have been crippled, when treatment was withheld so that the disease could be studied. More recently, it has been disclosed that the Army and CIA experimented with LSD on unsuspecting subjects in the mid-1960's. Extensive controversies have also arisen in recent years about the morality and ethics involved in psychosurgery and lobotomies performed on prisoners and mental patients. Throw in other experimentation disclosed in the last few years to have been performed on the poor, handicapped, mentally-ill, etc., and that which must have also taken place in foreign countries, and one almost has to wonder where it will all end. Will this whole situation become far more serious in years to come as the biological revolution unfolds? Should a set of testing ethics be drawn up today to prevent possible experimentation abuses tomorrow? Who should design them? What safeguards against misuse of the principles should be established?

How paradoxical and incongruent is it that man is currently (1) preparing to increasingly play God over the human species, while increasingly denying himself that role over other species; (2) gearing up to increasingly experiment on himself, while increasingly attempting to prohibit such action on other animals; and (3) increasingly experimenting on and endangering the lives of some helpless and handicapped people, while at the same time attempting to prolong the lives of others, as in the Karen Quinlan case?

Does man really have a sufficient understanding of the evolutionary process, and its origins, to attempt to control it? If man does decide to participate fully in designing and controlling both evolution in general and his own evolution in particular, can he then do anything else? Could he ever put that responsibility back upon natural Providence or God—without giving up being man?

6. HOW WILL THE BIOLOGICAL REVOLUTION AFFECT THE GOD CONCEPT AND A REVERENCE FOR LIFE AND NATURE? We already seem to be losing our sense of wonder, mystery, and awe today. We are seemingly becoming more impressed with ourselves and our accomplishments, and less impressed with God, life, and nature. We seem to increasingly think that we are writing the book of science and life rather than just reading it. And, in the process, we seem to be increasingly denying the possibility that there was another, original author. Won't the biological revolution simply aggravate that situation? For example, for many, it is the unacceptable, degrading, and even absurd nature of death that is one of the chief stimuli that prompts them to believe in God. But if we were to cease to die, that stimulus would be removed. If we become "king of the mountain," often pushing God off, and we no longer feel restrained by a sense of allegiance and responsibility to God for the Creation, what will restrain us in our relationship with nature and other men? Will our accelerating control of evolution, and our physical and mental improvement (perhaps eventually to even "superman" status), increase or lessen our regard for nature? Will it increase or lessen our dictatorial stance toward nature and our feeling of being the "superior animal"—unaccountable to nature and/or God? Can we afford a further weakening of the God concept and a lessening reverence for life and nature?

7. TO WHAT EXTENT WILL A WEAKENING OF THE GOD CONCEPT AND A LESSENING REVERENCE FOR LIFE AND NATURE AFFECT ORGANIZED RELIGION? On the eve of the Fourth World Synod of Bishops in Rome in 1974, Pope Paul VI stunned a Vatican audience with an ominous prediction. "The church is in difficulty," he said. "The church seems destined to die." There are many, of course, who would say that was long overdue good news. But is it? Almost every species other than man is instinctively more loyal and less aggressive to its own kind, and to future generations of its own kind. Within the species of man, if organized religion is not the leading spokesman for mankind, humanity, and future generations, who is? If the churches are not society's chief custodian and guardian of its long-range values and goals, who is? As futurist Jay Forrester has said, "As

enduring values are gradually perceived by a society, those values are cast into religious codes. The religious codes serve to freeze and to propagate the long-term values."[9] What could happen if organized religion were to dramatically decline in scope or even die? Forrester has provided a rather succinct answer: "Social systems tend to decay if a collapse in their long-term goal structures occurs. As the enduring values erode, emphasis shifts to short-term objectives. As the present is emphasized over the future, the result is long-term deterioration and further emphasis on the short run. As the goals decline, the decision processes change, and a downward spiral begins."[10]

If organized religion no longer speaks clearly and effectively for future generations, who will? Where will the answers come from to the following typical questions which the biological revolution will increasingly prompt? Does one generation have the "right," wisdom, or virtue to decide the fate of another? Is it morally "right" for the living to foreclose, some, many, or even all of the unborn's options, without their having any say in the matter? Could man become a dehumanized puppet if his choice, memory, and intelligence patterns are pre-designed and pre-fabricated before his birth?

If organized religion no longer speaks clearly and effectively for mankind and humanity, who will? If no one does, can the eventual death of humanity be prevented?

8. HOW WILL THE BIOLOGICAL REVOLUTION AFFECT MAN'S OTHER HISTORICAL INSTITUTIONS? Will the biological revolution produce a social and moral revolution which will not only affect religion, but some of man's other cherished institutions as well, such as marriage and the family? Loneliness and alienation are already rampant in our society today. What would be the effects of even more nebulous marriage, family, and interpersonal relationships? What will happen when it *literally* becomes difficult to determine whom we belong to and who belongs to us? Will people ask as never before, "Who am I?" "What am I?"

9. ARE SCIENCE AND TECHNOLOGY OUR NEW RELIGION AND SCIENTISTS AND TECHNOLOGISTS OUR NEW PRIESTS? It's an almost amazing situation! There still aren't any comprehensive conceptions available of man, God, nature, the evolutionary process, or their interrelationships. And, the cautionary voices which have traditionally given us some direction in questions dealing with humanity, mankind, and future generations are now being increasingly stilled. And yet, we seem to be willing to listen, and to agree almost without question and hesitation, when science and technology tell us that it is our destiny to now proceed with the control and alteration of the evolutionary process (and even alter man himself?). Why? The answer, of course, lies in our almost overwhelming faith and confidence in science, technology, and "progress."

That faith and confidence have grown so strong that, in a sense, it might be said that science and technology have indeed become our new religion, and scientists and technologists our new priests. Anthony Standen, in his 1950 book, *Science Is a Sacred Cow,* explains how this situation came about.

> Laymen see the prodigious things that science have done, and they are impressed and overawed. . . . Science has achieved so many things, and has been right so many times that it is hard to believe that it can be wrong in anything. . . . The benefits we have received from it are tremendous, all the way from television to penicillin, and there is no reason to suppose that they will stop. . . . Mere laymen, their imaginations stupified by these wonders are duly humble, and regard the scientists as lofty and impeccable human beings. . . . Since it is only human nature to accept such flattery, the scientists accept the laymen's opinion about themselves. The laymen, on the other hand, get their information about scientists from the scientists, and so the whole thing goes round and round like a whip at Coney Island.[11]

As a sidelight, it might be pointed out that since Standen wrote those words in 1950, about 95 percent of all the technological progress made in history has occurred, scientific knowledge has about quadrupled, man has walked on the moon, and machines have been landed on Mars! He goes on to say in his book,

> When a . . . scientist . . . makes some pronouncement for the general public, he may not be understood, but at least he is certain to be believed. No one ever doubts what is said by a scientist. . . . Scientists . . . have a monopoly on the formula "It has been scientifically proved," which rules out all possibility of disagreement. Thus the world is divided into scientists, who practice the art of infallibility, and non-scientists, sometimes contemptuously called "laymen," who are taken in by it.[12]

Standen very cleverly used tongue-in-cheek humor in his book in an attempt to make the reality he was attempting to describe a little more palatable. His remarks set the stage nicely for the author's contention that *all* men now seem to be afflicted, to at least some degree, with four "hangups" or syndromes. Standen has already described what might be called the "Infallibility Syndrome"—the belief that given enough time, money, and manpower, science and technology could solve any problem. It's only a short step mentally from that belief to what might be called the "Cavalry-at-the-Last-Minute Syndrome"—the belief that no matter how bleak or black a situation now becomes, science and technology will save us, even if they have to come "riding through the pass" at the last second. The belief that if something can be done it must be done might be termed the "Everest Syndrome." The name can be traced to explorer George Mallory, who was once asked, "Why do you want to climb Mt. Everest?" His answer became the now almost tedious cliche, "Because it's there!" The Everest Syndrome surfaced almost as purely in an editorial in the July 14, 1969 issue of *Chemical and Engineering News,* which concluded by remarking, "That's about as good a reply as any to those who ask why a lunar manned landing." It's only a short mental step from the Everest Syndrome to what might be called the "Thank God Syndrome." This syndrome involves proceeding with a major project before there is a consensus as to what the exact consequences will be. We seem to be almost compulsively driven at times to do things—listening not to a minority of scientists who warn against proceeding, but the majority who assure us there is really nothing to worry about. It almost seems like we play Russian roulette with nature at times today! When the empty chamber comes up, however, and the uncertainty is resolved, we all breathe a sign of relief and exclaim, "Thank God!"

At times today, it seems like science and technology are in the saddle and ride us instead of the other way around. It's difficult to say to what extent the four syndromes just mentioned actually motivate us, for outwardly we usually justify our actions by saying certain things must be done for the sake of "progress," national pride, and to satisfy our instinctual "need to know" and quest for knowledge. Such arguments have already been used to advance the biological revolution and will probably be increasingly used in the future.

Challenging the American philosophy of unlimited "growth" and "progress" can result in one being accused of being unpatriotic, anti-American, anti-science, anti-technology, etc. Are there some things, however, that mankind might be better off not knowing and/or not doing? And, even if all knowledge is good, are there things that we shouldn't do with that knowledge? Where does our quest for knowledge and "progress" fit into the evolutionary scheme? Into a relationship with God? Is "progress for the sake of progress" and "knowledge for the sake of knowledge" morally defensible? Is restraining "progress" and knowledge in any way, morally defensible? Such questions are far more profound, and far more important than they appear at first glance!

10. IS OUR NEW RELIGION (SCIENCE) FLAWED? Some believe there are 3 major limitations to the scientific method, and thus to science itself, although most laymen are apparently unaware of that fact and most scientists either never realize, acknowledge, or discuss it. First of all, although the scientific method consists of several steps that emphasize experimentation, it is also comprised of several summarization and explanation steps which emphasize thinking. The laws, hypotheses, and theories which result in those latter steps are thus the product of the human mind. Since true scientists know that thought can be untrustworthy, they are constantly on the alert for that one little fact that can "upset the applecart." They also know that every measuring device they may use has limitations that limit its accuracy to some degree. And thus true scientists will never speak of completely proving or verifying a theory, but will instead simply indicate that they have a certain

degree of confidence or trust in it, and that there's a certain degree of probability that the theory is correct. As the nineteenth century scientist, J. J. Thomson, once said, "A theory is a tool, not a creed."

Secondly, science cannot solve all problems because the scientific method is limited in its effectiveness. It works most impressively when it is used to understand and predict the behavior of individual atoms and molecules. But as molecules get more and more complex, their behavior usually becomes less and less predictable. If the roles and actions of a series of such molecules are interwoven, their collective behavior becomes even more difficult to explain. And, if those interrelated molecules comprise a living thing, the overall behavior of that thing is even less predictable yet, because there is an overeffect superimposed on that collection of molecules of which it is comprised—life! Scientist John Rowland stated all of this a little more succinctly over a decade ago when he said, "As science moves from physics, through biology and psychology, to sociology, there is to be observed a process of increasing abstraction. The more concrete assumptions of physics are gradually replaced by abstractions not easy to comprehend. Atoms are replaced by ideas like genes and chromosomes."[13]

Science does not understand life and perhaps never will! British philosopher and scientist Michael Polyani has said, "When I say that life transcends physics and chemistry I mean that biology cannot explain life in our age by the current workings of physical and chemical laws. . . . Such a highly improbable arrangement of particles is not shaped by the forces of physics and chemistry. It constitutes a boundary condition, which as such transcends the laws of physics and chemistry."[14] Many would argue, of course, that Polyani's contention is overstated and that the situation he describes may change. However, if he's right, it seems safe to say we may never understand the behavior of the most complex collection of molecules of all—the human brain. Individual atoms and molecules behave in largely predictable ways, but the mind usually does not! And thus it is that chemistry and physics are well developed sciences, but psychology and psychiatry are not.

Thirdly, because the scientific method has not proven to be a very effective tool in understanding and predicting man's behavior, science can really only promise us truth. It really has nothing definitive to say about the good or evil and thus desirability of that truth and whether or not it will bring us peace, love, and happiness. Those judgments must be made in the scientist's mind. However, since most scientists are experts on nature rather than on human nature, they are usually no more qualified to make those judgments than anyone else and should thus also usually be considered to be no better prophets than anyone else! As Albert Einstein once said, "Science can only ascertain what 'is,' but not what 'should be.' " Anthony Standen has taken that idea a little further by rather pointedly claiming that we are turning out overspecialized scientists today who know *how* to physically change the world but have very little idea of *what* to change it into.[15] In other words, they are long on knowledge, but short on wisdom!

11. IS OUR NEW PRIEST (THE SCIENTIST) ALSO FLAWED? The answer is yes, of course, because he or she is a human being! As Anthony Standen has said, laymen regard scientists as being "lofty and impeccable human beings." In fact, however, scientists are usually plagued by the same frailties and flaws that plague everyone else—sometimes, even more so!

First of all, there is at least some evidence to indicate that scientists are indeed "different."[16] On the average, they apparently are more introverted and less interested in social relations than non-scientists. They really do live at times within the walls of their proverbial ivory towers, where contributions to science can be merely an end in themselves. Science is the major arena, and at times, the only arena in which many scientists express their instinctual need to compete and excel. It is often the only area in life in which they attempt to "do their thing" and seek the power, importance, recognition, and fame which all humans seek to varying degrees. Sometimes the scientist's quest for power gets out of control, just as it does for the non-scientist. The late author C. S. Lewis warned of that danger more than a quarter of a century ago when he said, "Man's power over Nature is really the power of some men over other men, with Nature as their instrument."

Secondly, a scientist is automatically born on a pedestal—a pedestal whose base consists of 350 years of success stories. His own personal success, coupled with the added respect and adoration it may prompt, may raise that pedestal even higher. At this point, the scientist may be consciously or unconsciously changed by success and power, just as anyone else might. For example, the scientist may begin to place a greater and greater emphasis on personal, professional, and financial advancement. But he may also begin to lose his humility and begin to get "a little too big for his britches." He may dwell on his successes to the point where he begins to wonder if science really isn't infallible. At this point, his theories can then begin to imprison him rather than stimulate him. They can become creeds rather than tools and they can begin to hide the facts they fail to explain. The scientist may begin to lose his sense of wonder at this point, as he puts excessive emphasis on what he knows rather than what he doesn't know. As he becomes more impressed with certainty than probability, he may begin to find it virtually impossible to deny his philosophy in any way. And, when he reaches the point of being rather intellectually provincial, he's obviously not going to like "the boat rocked."

Thirdly, the scientist is at least as overspecialized as the non-scientist in most cases. He tends to work within a corner of a corner of a corner, whether it's in an ivory tower or not! The scientific movement is thus so broken up and scattered that only a few scientists have a grasp of what their discipline of science is doing overall, fewer still what science is doing overall, and only very seldom is there a scientist who is current on how science is fitting into a "big picture."

In summary, science is practiced by humans, using an instrument (the scientific method) that has severe limitations. For these and other reasons, *the scientist all too often suffers from the same lack of social and global concern, foresight, responsibility, and caution which afflicts many non-scientists.*

12. IS SOCIETY PLACING TOO MUCH TRUST AND CONFIDENCE IN SCIENCE? Should scientists be trusted and/or allowed to make decisions which could drastically affect all mankind in a planned or unplanned way? Or, are there actually too many conflicts of interest involved for science to ever really be able to make such decisions in a completely unbiased way? Can science self-regulate itself? If not, to what extent should the public and government be involved in the regulation process?

Although such questions are coming into sharper focus today, and are looming larger and larger, there are no clear-cut answers available to them yet. Unfortunately, the most dramatic case history which might be cited in support of that statement is also a current situation where such answers are needed the most.

The "gene cloning," plasmid, recombinant DNA techniques which were discussed earlier in chapter 2 of Part I were developed in the early 1970's at the Stanford University School of Medicine and the University School of Medicine in San Francisco. Realizing the potential dangers in such experiments, a committee of renowned scientists, headed by Paul Berg of Stanford, wrote an open letter in July of 1973 to the journals *Nature* and *Science.*[17] The letter called for an unprecedented, voluntary, self-imposed moratorium on such research. The moratorium went into effect almost immediately, probably because the letter and its request had the backing of the National Research Council of the National Academy of Sciences. In February of 1975, an international group of 140 scientists met at a four-day conference at Asilomar, California to decide whether the by then 18-month moratorium should be lifted, and, if so, whether a set of guidelines should then be established for subsequent research.[18] After a rather stormy conference, an almost last-minute-decision was made by majority vote to end the moratorium and accept a set of provisional, temporary guidelines which had been drawn up almost overnight! A committee of the National Institutes of Health (NIH) Advisory Committee then met several times in the next 15 months to draft more comprehensive guidelines. A long and boisterous public input session was held at Bethesda, Maryland in February of 1976. A final set of guidelines was published that June as the "National Institutes of Health Guidelines for Research Involving Recombinant DNA Molecules." The guidelines essentially spelled out how the risk of spreading foreign, recombinant DNA through the environment is to be minimized. Essentially, this is to be accomplished through the use of "P1 to P4" types of physical containment within the laboratories where such work is done and the development of "EK1 to EK3" types of biological containment, wherein only enfeebled and disabled host material is used which can only propagate under specialized laboratory conditions.

At first glance, the sequence of events just described appear to have come to a desirable and equitable conclusion. It would seem that, prompted by a concern for society, science was able to self-regulate itself, and that society had a substantial input and voice in that action. However, a series of disturbing questions in five broad areas remain unanswered, leaving lingering, nagging doubts as to just how desirable and equitable that conclusion really was.

First of all, simply holding the Asilomar Conference in itself was evidence that by February of 1975 most scientists had already decided that recombinant DNA research should proceed. The extent to which that decision was actually prompted by the "Everest Syndrome" and the promise of such short-term benefits as the Nobel Prize and other professional rewards, fame, job tenure, wage increases, etc., will obviously never be known. The major question to be resolved at the conference was thus really whether guidelines were needed for such research, and, if so, what they should be. The argument seemed to largely center on the question of the possible dangers and abuse of such research. But one can't help but wonder to what extent expediency once again prompted a decision. To what extent was it actually concern for society that decided the question, and to what extent simply a fear of possible governmental regulation of academic freedom, research, "progress," and "doing one's thing"?

Secondly, although the public will essentially pay for much or even most of the recombinant DNA research to be done through taxes funneled through the NIH and other governmental agencies, did it really have the input and participation in the decisions made that it should have had? Or, did it merely play a token advisory role? Why was public comment and participation solicited only after scientists had already written draft versions of the guidelines?

Thirdly, exactly how dangerous are such experiments? Proponents of recombinant DNA research usually emphasize their belief that nature rather routinely does recombinant DNA experiments of her own. But they do not emphasize the fact that Marberg virus, green monkey disease, and "legionnaire's disease" could be deadly recent examples of the fact that her experiments are not always successful.[19] Can the extent of the danger involved in scientific experiments really be fully predicted before the experiments are done? Do we really know enough about DNA, and its control and expression mechanisms, for example, to say for sure that there won't be some major surprises and unexpected results in such research? Obviously we do not, for the scientists involved in such research invariably use the words "we believe" rather than "we know" when they speak of the risks involved, and they always use the words "minimal danger" rather than "no danger." In fact, use of the words "risk" and "low, medium, and high containment" in the NIH guidelines indicates in itself at least some measure of potential danger being involved. In September of 1974, Sir John Kendrow, president of the British Association for the Advancement of Science, said publicly, "The consequences of developments in nuclear physics were easier to predict in 1939 than are the possible consequences of gene transfer in 1974."

There is an old folk saying that disaster can be anticipated in varying degrees from appropriate combinations of "kids and fools and sharp-edged tools." To what degree must the danger created by the possibility of human error and accidents be added to the danger of surprise and unexpected results just mentioned? Potentially serious accidents have already occurred in high containment laboratories. For example, despite extreme contamination precautions, samples of moon dust escaped through a defective glove box after the first lunar landing. On a later flight, a bacteria was allowed to reach the moon and return. In the fall of 1976, frantic but successful isolation efforts involving 45 people possibly prevented global tragedy after a British scientist working with green monkey disease in a high security laboratory accidentally punctured his protective glove with a hypodermic needle.[20] If altered, dangerous *E. coli* were to escape from a high containment laboratory, perhaps as simply as through an experimenter unknowingly breathing or ingesting some of the material, and carrying it out in his body, there would be no way of determining that an accident had occurred until *after* the organism had begun to do its irreversible damage. As Dr. Robert Sinsheimer, a biologist at the California Institute of Technology, said recently, "Because of human fallibility, these new organisms are almost certain to escape. There's no way to recapture them and thus we have the great potential for a major calamity."[21] Wouldn't the danger of such an accident increase as the number of laboratories doing such research increases?

The danger level rises even higher if one figures in the possibilities of natural disasters, sabotage, terrorism and blackmail, the "mad scientist," and the development of genetic weapons research. And what of the scientists with lesser abilities and qualifications who are attracted to the work but cannot afford the precautions needed? Or the scientists that have such overweening confidence in their ability that they feel the regulations and precautions are unnecessary, and only so much red tape. Interestingly enough, all of these possibilities were rather carefully avoided in discussion at the Asilomar Conference.

Fourthly, are the current guidelines enforceable? Without the power of law behind them, what power do they have? Can they be enforced through only peer pressure and the honor code? Would industry comply with a voluntary or even legal registration and inspection system in which it could lose its competitive edge? Would it be fair to legally force American researchers to comply if researchers in other countries were not regulated? How would global regulation be accomplished? How will compliance with the guidelines be accomplished when researchers with less abilities and qualifications realize that "exciting" recombinant DNA research can be done in almost any laboratory that can handle pure bacteria cultures—perhaps even in high school laboratories? Can curiosity and temptation really be completely controlled on a voluntary basis? On a legal basis?

Fifthly, one can't help but wonder whether mankind, humanity, and future generations were adequately represented in the decisions to proceed with recombinant research that have already been made. Recombinant DNA research has been largely justified with the claim that it will eventually lead to the alleviation or even elimination of genetic disease. However, there has been very little discussion of the more distant possible ramifications, such as other aspects of negative eugenics, positive eugenics, and the removal of various roadblocks standing in the way of other aspects of the biological revolution. To what extent has recombinant DNA research been caught in the same problem that affects us all—a short-term perspective? Are we once again caught in the trap of being too concerned about today, and forgetting that tomorrow will also come?

The NIH guidelines for recombinant DNA research were introduced in June of 1976. In the following year, the five areas of problems just discussed, "loose ends" in a sense, prompted an increasingly noisy, impolite, and polarized debate in the scientific community.[22] Increasing concern and opposition spilled over into the public and governmental realms by early 1977, to the extent that on February 4 a bill was introduced in Congress requiring compulsory licensing of recombinant DNA research, with civil and criminal penalties for violators.[23]

The French statesman Georges Clemenceau once said, "War is much too important a problem to be left to generals." Have we now reached a point in history where society is now beginning to say similar things about the role of science and scientists in life and the future?

13. SHOULD A SCIENTIST BE HELD RESPONSIBLE FOR THE CONSEQUENCES OF HIS WORK? Many scientists would unhesitatingly answer that question *no.* However, many of the arguments they would offer in support of that position are actually only rationalizations, and are probably at least as infantile as they are valid. For example, statements such as "If I hadn't done it, someone else would have," "It was my duty," "Theoretically, it should have worked," etc., could be so labeled. But there is one argument that is much more sophisticated and complex than the others. It's the idea that

once a discovery is made, it's up to society to decide how to use it. Or, as some have said, "Science may provide the means, but men choose the end." Many scientists actually believe that since there are no forces or objects in nature that are "bad" in themselves, and since science, the scientific method, and the pursuit of truth are neutral to the ideas of good and evil, their discoveries are neutral as well. And thus, a sort of "wash my hands of the whole thing" philosophy actually does exist in science. Many scientists really do feel that it is unreasonable and even paradoxical to expect a scientist to make a discovery and then forecast its consequences of even lobby for controls of its uses.

In hearings held before the Senate Subcommittee on Government Research, held in March of 1968, Nobel Prize-winning geneticists Arthur Kornberg and Joshua Lederberg testified that geneticists should not be asked to make critical judgments on the legal and ethical questions arising from genetic research and development. Kornberg told the subcommittee that "there are absolutely no scientific rewards in terms of furthering scientific skill by involving oneself in public issues. Scientists deal with molecules, impersonal objects, and you can get rusty quickly. The more I am out of that area, the less competent I am in that field." Kornberg did add, however, that the public should be educated about genetics research so that it can participate in the questions that will need answers. As one listens to such arguments, one can't help but wonder to what extent humorist Tom Lehrer was actually describing reality a few years ago when he said,

> "Once the rockets are up
> Who cares where they come down
> That's not my department."
> Says Wernher von Braun.

Where is the balance point between a scientist's responsibility for himself and his responsibility for and to all humans? Can a scientist really be indifferent to the results of his research? Or, is he a human being and a member of society and mankind as well as being a scientist? Is he, like all other men, his brother's keeper? As a member of mankind must he be concerned with what *we* ought to do over and above what *he* can do? In fact, does the scope and importance of his contributions to society and mankind make him even more responsible than most for what he does? Aren't there almost countless lessons in history which indicate that a scientist should automatically suspect and perhaps even expect that his discoveries will be misused and even abused, and plan accordingly? And if the gap between science and ethics, and spirituality really is widening today, shouldn't the scientist at least consider the possibility of "the smash" which Edison warned of many years ago? While it is true that loaded guns are not "bad" in and of themselves, can they really be placed in the hands of children and not eventually cause serious trouble?

It might be wise for scientists who believe they have no responsibility for their discoveries to review what happened in Germany in the 1930's and 1940's. "The German universities expressed less opposition to the growth of Nazism in the 1930's than did the churches and labor unions; their failure at this point has been attributed in large part to their neutral pursuit of truth without concern for the life of the nation."[24]

Doctors, lawyers, large companies, etc., are being sued today for their mistakes and lack of concern and responsibility for individual men, mankind, and the future. Could the day come when scientists will be sued in the same way? Is it even possible the day could come again when scientists will again sit before a Nuremberg tribunal and perhaps even be hung?

14. IF THE SCIENTIST IS NOT RESPONSIBLE FOR THE USE OF HIS DISCOVERIES, WHO IS? If the scientist is unwilling and/or unable to assume responsibility for his discoveries and/or self-regulate them, is the laymen willing and prepared to assume that role? Should he? Who else is there? What is the answer to the question which Francis Ensley asked over a decade ago—"Science has enabled man to control Nature—but who's to control the controller?"[25]

15. WHAT ARE THE RESPONSIBILITIES OF THE RELIGIOUS PERSON AND ORGANIZED RELIGION IN THE SERIOUS AND EVEN CRUCIAL QUESTIONS WHICH HAVE NOW BEEN RAISED? IN THE BIOLOGICAL REVOLUTION AS A WHOLE? Unfortunately there can be no answers to such questions unless a rather complex prerequisite question is answered first. That prerequisite question must be prefaced with a short history lesson.

It would *seem* to be clear from the Gospels that Christ was concerned for the welfare of *both* the souls and bodies of men. However, the epistles and letters of John and Paul, and other writings, cautioned early Christians not to love the world, or worldly things, and to stay aloof from political and secular societies. As might be expected, this led to a subsequent, heavy emphasis on "saving souls," "individual salvation," and a general historical trend toward isolationism. As contemporary theologian H. Richard Niebuhr has said, "Social responsibility has been somewhat obscured during the long centuries of individualistic overemphasis."

However, a battle developed in the 1960's, for various reasons, in many churches between "liberals" who often seemed to almost want to convert the church into a social welfare agency and "conservatives" who cared a lot about saving souls but very little, it seemed at times, about feeding hungry babies or other worldly concerns. The argument never really got around to its theological basis—the relationship between faith and good works—but soon simply emotionally polarized into an "either-or" choice. Some people cut their pledges because they disapproved of church involvement in social concerns. Others shut their pocketbooks on the grounds churches were not doing enough to promote human welfare. Since the 1960's, social and political activism in the churches has waned and there has been a gradual trend back to the traditional doctrine that emphasizes the role of faith in personal salvation. Overall, it would appear that the "conservative" branches of organized religion came out of the 1960's less "scarred" in terms of demoralization, attendance, and giving than the "liberal" branches. The "liberal" branches appear to be in a retrenching stage in the 1970's and appear to be generally operating now from a less "liberal" stance.

The battle is still not over, however, for the basic question at its heart is yet to be resolved. Evidence that the battle is still at least simmering can be seen in both the so-called "Hartford Appeal," a white paper published by 18 "conservative" churchmen in January of 1975 which attacked the church's "surrender to secularism," and the subsequent "Boston Affirmations," a paper published by 21 "liberal" churchmen in January of 1976, which urged a reaffirmation of the church's active role in the world. Should the major concern of the religious person and organized religion be personal piety and salvation, or social concern and action? *Or, should it actually be both? Does the church have a responsibility not only for man, but for his society as well?* To what extent was Marcus Aurelius right when he said, "What is not good for the swarm is not good for the bee"? How would one refute the words of theologian H. Richard Niebuhr when he said over a decade ago,

> If a man responds to the demands of a universal God, then the neighbors for whom he is responsible are not only the members of the nation to which he belongs but the members of the total society over which God presides. . . . The church cannot be responsible to God for men without becoming responsible for their societies. . . . If the individual sheep is to be protected, the flock must be guarded.

16. IS RATIONALITY RATIONAL? The last general question to be asked may be the most disturbing of all. Are there really ageless, time-proven truths, or is truth actually to be found only in the mind of the beholder? If there is a God, it would seem that reason, responsibility, and rationality have at least a fighting chance of being rational. But if there is no overall blueprint or master plan in operation in history, where is the anchor for holding reason, responsibility, and rationality solidly to life? If there is no God, who set up the rules for the game, and who now serves as a referee, the game is open! Life is simply an accident and evolution is a random, even blind process with no defined goal(s). All truth would be relative in that situation, and thus would exist only in the mind of the beholder. There really would be no solid, binding answers to turn to in the area of morality and ethics, or any other area as well. "Indicative" or situational ethics (and actions) would thus reign supreme.

There would really be only one conclusion that one could consider to be final—that there are no final conclusions! Such a situation suggests a famous passage in the Chandogya Upanishad in which a father instructs his son about the nature of reality through various actions like peeling an onion, or dissolving a grain of salt in water. As layer after layer is removed, the boy finally reaches a point at which he finds nothing at all. There is a void at the heart of the onion, and the salt grain disappears as the last layer dissolves. Yet the father explains that the invisible central point is the ultimate essence, not only of the phenomenal universe, but of the boy himself. The boy is truth!

The question of whether rationality is rational is perhaps the biggest, as well as the most disturbing question of all about the biological revolution. For if rationality is not rational, man has the *full* responsibility for his survival or his extinction fully on his own shoulders. The decisions man makes about the biological revolution would be critical and would have to be "letter-perfect" because of their irreversibility. There would be no margin for error! In the blind randomness of nature and evolution, humans would exist only by historical and geological consent, and that consent could be revoked without notice!

If ultimate and final truth resides only in the mind of the beholder, humanity is in the paradoxical situation wherein to accomplish anything collectively we must be skeptical of all individual goals and accomplishments! If that is the case, we had better hope (and even pray) that biologist Leon Kass was wrong when he recently said, "Contrary to popular belief, we are not even on the right road toward a rational understanding of and rational control over human nature and human life. It is indeed the height of irrationality triumphantly to pursue rationalized technique, while at the same time insisting that questions of ends, values, and purposes lie beyond rational discourse."[26]

MISCELLANEOUS OVEREFFECTS

The 16 overeffects just discussed are general overeffects which the biological revolution could produce. There will obviously be important specific overeffects produced as well, which also deserve discussion. Unfortunately, however, space considerations won't allow more than passing mention of several random, representative examples in this respect.

1. What are the moral and ethical implications of creating the groups of people who are physically and mentally superior to the rest of the population? Do we have the "right" to create a society where all men are not created equal?
2. Could the survival of capitalism, communism, or socialism depend on which form of government adjusts best to the biological revolution?
3. What are the moral and ethical implications of producing "happiness" chemically and/or electrically? If everyone is "happy" can anything be "wrong" or "sinful"?

4. Dr. Hans Jonas of the University of California maintains that genetic alteration research is "ethically unsound." "At least one try at real cloning or at really producing a genetically altered individual is necessary to find out what is possible or what the achieved possibility really is like. The very deed eventually to be decided on in the light of knowledge already is committed in the night of ignorance in obtaining that knowledge."[27] Dr. Jonas has gone on to say that since only the scientist performing such an experiment can really be "informed" about the possible consequences, it is he or she who should be the first to volunteer!
5. Creation of life research could indicate the possibility of extraterrestrial life so strongly that science might be convinced to undertake a massive search for such life. What are the dangers of a manned landing of Mars or a space probe bringing back a virus or microorganism for which humans have no defense? Suppose we found intelligent extraterrestrial life, or it found us? What psychological ramifications would be involved? What would a sudden confirmation that we are not alone and not the center of the universe do to us? What would the discovery that we are actually an inferior species in the universe do to us? Would we find ourselves without dreams, our horizons closed, our intellectual and spiritual aspirations as outmoded as to leave us paralyzed? How would we interpret God's plan of salvation for us earthlings—the mission of Christ in light of this new information? Are we really prepared for the possibility of cosmic company, even if it's friendly? If it's unfriendly or even hostile?

IN SUMMARY

In chapter one we raised the question of whether man is building a modern Babel. After seeing some of the possible primary implications and societal consequences of the biological revolution outlined in chapters two and three, and the possible secondary overeffects outlined in this chapter, the same two questions asked previously almost beg to be asked again. First of all, are we actually building a Bio-Babel? And secondly, *how great a rate and degree of change, complexity, and future shock can society stand?* Will society be resiliant enough to absorb the added psychological stress which the biological revolution will add to that which is already present and that which is coming automatically as the future unfolds? Added to the continuing psychological stress produced by the revolutions begun by Galileo, Darwin, Freud, and others; the current and coming physical and mental stress produced by the sexual, urban, computer, pollution, population, food, weather modification, resource, and energy crises; the current and coming psychological stress produced by modern weapons; and the possible psychological shock produced should humans ever discover extraterrestrial life; could the biological revolution prove to be the "last straw," psychologically speaking? Individuals have psychological "breaking points." Does society also have one?

The spectre of Bio-Babel raises another critical question in addition to our possible resiliancy. That question is simply whether or not humans are intelligent enough and sophisticated enough to deal with questions of a nature never asked before in history, questions with the most complex moral, ethical, and spiritual flavor ever encountered in history. There are some who believe that the modern man and woman are not equal to the task and may already be "obsolete"!

> The ethical problems . . . raised by the population explosion and artificial insemination, by genetics and neurophysiology, and by the social and mental sciences are at least as great as those arising from atomic energy and the H-bomb.[28]
>
> —W. H. Thorpe

> It represents probably the largest ethical problem science has ever had to face. I fear for the future of science as we have known it, for mankind, for life on the Earth.[29] —George Wald

> It is as if the biochemical and bioenergetic facts, potentialities we are beginning to elucidate, were waiting in ambush for man. It may well prove to be that the dilemmas and possibilities of action they will pose are outside morality and beyond the ordinary grasp of the human intellect.[30] —George Steiner

> Do we really face an ambush? Or is this an epic opportunity? Is this a dilemma too dense to penetrate, a potential too large to grasp? Or is this the goal toward which evolution has been striving for five billion years: to be in its product aware of its essence and thus to rise above chance?[31] —Robert L. Sinsheimer

> Western civilized man is a Stone Age organism trying to exercise 21st century power in a world of 18th century institutions, based on medieval humanistic principles.[32] —John S. Lambert

CHAPTER 5

BIO-ACTION

> It's not enough that we do our best: sometimes we have to do what's required. —Winston Churchill

If we really are in the midst of building a Bio-Babel, as the author contends, we could also be painting ourselves into a historical corner—a corner from which we will leave "tracks" no matter how we emerge. In other words, having seen the extremely complex questions which have now been posed about the biological revolution, one might suspect that the question of what to do about it might be extremely complex also. Unfortunately, it is! Although it is easy to list the action alternatives possible, it is extremely difficult, at this point in history, to decide which one(s) to follow.

POSSIBLE ACTION ALTERNATIVES

It is difficult to envision any more than the following three alternatives or options as possible action responses to the biological revolution:

1. Ignore, accept, or endorse it.
2. Attempt to control or slow its pace, overall or in part.
3. Attempt to prohibit it, overall or in part.

Ignoring is listed as part of option one, for the simple reason that disregarding the biological revolution will produce essentially the same results as accepting or even endorsing it. It should also be pointed out that, whereas option one will always be available, options two and three are going to be increasingly difficult to initiate as time goes on. Although it's difficult to say exactly how long it will be before "the cat's totally out of the bag," it should be obvious that such a time may not be too far off.

Ignoring would not seem to be a viable option for the concerned, responsible person. It's also difficult to believe that such a person could simply accept or even endorse the biological revolution in its entirety, for that would necessitate completely dismissing all of the concerns that have been raised previously in this book. It would also require assuming that we are masters of our fate, that society is completely under control, and that our historical course is set and that we are not adrift, being carried along by forces we do not have the courage or the strength to master. Such blanket assumptions have to be at least somewhat uncomfortable intellectually today.

There is a distinct need for some national polls to be taken on various aspects of the biological revolution. The author has done some polling in this respect, on a small scale, with several hundred students in his classes from 1971 through the fall of 1976. The results are shown below. In the column labeled "With Thought," the 256 students polled (in a science course for non-science majors) were allowed to think about the six developments listed for about a month before voting as to whether or not they thought such research should proceed. The "Spur-of-the-Moment" students voted immediately.

RESULTS OF STUDENT POLLS ON THE BIOLOGICAL REVOLUTION

	Percentage of Students Voting Endorsement		
Development	With Thought (256 students)	Spur-of-the-Moment (119 students)	Overall (375 students)
Memory and Intelligence Modification	69	55	62
Development of an Aging Drug	46	56	50
Genetic Engineering	62	59	60
A Genetic Determination of Possible Racial Superiority	20	—	20
Cloning Humans	15	17	16
The Creation of Life	30	31	30

The above data obviously shouldn't be taken too seriously. However, there are six general observations that can probably be safely drawn from it. First of all, it seemed to make very little difference in the results whether the students voted immediately or after considerable thought. This would seem to indicate that the six developments listed all evoke some fairly strong initial feelings and emotions. Secondly, the voting pattern was rather consistent through the years, with no dramatic differences in the vote from class to class. Thirdly, the strongest endorsement was 69 percent, but for memory and intelligence modification, an area in which one might expect considerable student interest. Fourthly, it would seem to be ambiguous to vote for genetic engineering and then oppose using the same knowledge and techniques to study racial differences and carry out cloning. Fifthly, the "mini-poll" would seem to be at least weak evidence that most people would not simply accept or endorse the biological revolution in its entirety once they realize what it entails. And lastly, there was some indication that older students voted more negatively than younger students.

Of those students who believed research should not proceed on one or more of the six developments, almost half believed such research should be prohibited by law. The idea that there are things that humans shouldn't know and/or shouldn't do with their knowledge seems to slowly be gaining acceptance today. For example, Herman Kahn, writing in his book of some years ago, *Can Man Survive the Future?*, urged the establishment of an index of forbidden knowledge. But isn't prohibition of knowledge probably as unrealistic as blanket endorsement? Wouldn't attempts at prohibition be extensively criticized as being "anti-progress," "anti-science," and "anti-research"? How long would it take to marshal enough public, governmental, and scientific support to initiate such attempts? If such attempts were initiated, who should determine what research (and in what areas) should be prohibited and what laws in that respect should be written? If advisory commissions or governmental agencies draw up the laws, who should staff them? Should the public be given the right to vote on such laws in general referendums? Does increasing financial support of scientific research give the public the right to regulate science? Does the general public actually have the knowledge, ability, and expertise needed to do it intelligently? Should it be Congress that decides on the laws? Could they do it intelligently? If such laws were actually passed, how would they be enforced? What should the penalties be for breaking them?

Although the United States decided not to develop the SST, England, France, and Russia did. Of what use would it be to prohibit the biological revolution, or parts thereof, in the United States if other countries were to go ahead with such research? What would prevent scientists in this country from simply going elsewhere to continue their research? If such research continues and becomes widespread around the world, is there a distinct danger that it could become irresponsible at certain times and in certain places? How would growing global differences in affluence influence the development of such research? Could a developed country justify providing itself with intelligence enhancement, genetic engineering, life prolongation, etc., when elsewhere people are dying at early ages because of a lack of technology and its benefits? Would developed countries actually have a responsibility to make such techniques as intelligence enhancement available to underdeveloped countries in order to help them "catch up"?

It should be obvious from what has just been said that the biological revolution could create problems too large to be solved on anything but a global basis. Do we need an international commission to consider what should be done about the biological revolution and its possible consequences—including the possibility of secret warfare? Who should be on that commission? To whom should the commission report and be responsible? Will we need biological revolution non-proliferation treaties, just as there is a nuclear non-proliferation treaty? What other types of international agreements might we need? What roadblocks to such agreements will result from differences in outlook on reproduction, population levels, sex, marriage, and even life itself around the world? How would any recommendations, agreements, or treaties drawn up be enforced? Lord Ritchie Calder, a British science expert, has suggested we will need "an international police force of scientists, doctors, politicians, and other people."[1] He believes they should "have the power to inspect laboratories and enforce guidelines on the kinds of research that are to be allowed," and that "the guidelines and the powers of this police force would have to be negotiated among the nations involved." Is Lord Calder's plan workable? Or, would we need the actuality of a world government before the biological revolution could be effectively controlled on a global basis? Could that be too late?

If it is undesirable to ignore the biological revolution, unrealistic to either completely accept or endorse it, and unrealistic to attempt to prohibit it overall or in part, there is really only one alternative left. That, of course, is option two of the three listed previously—to control or slow it down, overall or in part. Biologist Leon Kass summarized the need for this approach in the area of genetic engineering when he said in 1971, "When we lack sufficient wisdom to do, wisdom consists in not doing. Caution, restraint, delay, abstention are what this second-best (and perhaps only) wisdom dictates with respect to the technology for human engineering."[2] Bernard T. Feld, editor of the *Bulletin of Atomic Scientists,* issued the same warning in 1976, but much more succinctly, when he said, "When you don't know, GO SLOW."[3]

There are probably only two major ways in which the pace of the biological revolution, or parts thereof, might be controlled or slowed, however. The first is simply self-regulation by science itself. While it's true, however, that some scientists exercise caution in or even abstain from research for which they have severe reservations or fears, all too many do not. In fact, although many scientists profess concern about the societal consequences of science, only rarely is there one who doesn't balk at the suggestion that it be regulated. It's probably thus just not realistic to expect science and individual scientists to both advance knowledge and simultaneously effectively control it! This was the ambiguous role which the Atomic Energy Commission attempted to play for many years—a role for which they were continuously criticized.

The second way would be to give the Office of Technology Assessment or a Science Court, if one were established, the legal power to control or slow the pace of the biological revolution by specifying the length of certain testing periods needed for certain developments, requiring the filing of "societal impact statements" similar to the environmental impact statements required today, determining the overall chronology to be followed in each development, etc. Perhaps what we actually need is an Office of Biological Administration similar to the Federal Drug Administration (FDA) in existence today. Such an "OBA" could perhaps evaluate, test, license, and regulate biological developments, just as the FDA deals with chemicals and drugs today.

There is, of course, no Science Court or OBA in existence yet or even "on the drawing boards." And, furthermore, the Office of Technology Assessment is currently so weak that for all practical purposes it's essentially non-existent. Thus it is probably also not realistic to believe that the pace of the biological revolution will be controlled or slowed by governmental regulation in either the immediate or even near future. What is left, then, if both scientific self-regulation or governmental regulation appear to be infeasible at the moment and are, in fact, nowhere on the horizon?

At some point in this chapter, the author is obviously going to have to unveil his own "game plan" for dealing with the biological revolution. Perhaps it's time to do that. The author believes, as the reader may already have suspected, that we are not yet to a point in history where science, research, and "progress" can be prohibited. He is sympathetic, however, with those who may disagree and feel inclined to try that option! But, although the author believes that the biological revolution will proceed largely unchecked in the immediate future, he does advocate initiating efforts to control or slow its pace —at least in part. Unfortunately, however, it appears that the government will not effectively regulate the biological revolution, and science will not effectively regulate itself, until one or both are sufficiently pressured or even forced by the public to do so. And thus the author advocates *group* pressure in this respect. Such pressure by individuals may just not be enough.

The only "power blocs" left in society which can probably still exert such pressure are educational groups, organized religion, environmental groups, the League of Women Voters, etc. Such a group or power bloc might conceivably call for governmental control or self-regulation of certain aspects of the biological revolution. However, the group might be able to exert even more pressure by issuing statements, resolutions, and petitions, and/or using lobbyists to call for a moratorium on certain aspects of biological research. Obviously it could take a great deal of "homework," discussion, debate, and even argument before the group might decide which parts of the biological revolution should be included in the moratorium, and how long the moratorium should be. But, no matter what the group decides to do, there is a "trump card" which it can play in ways limited only by the group's imagination. The biological revolution has a soft spot, and a vulnerable underbelly. It is currently being financed in large part through the National Institutes of Health, and other governmental agencies, *with tax money!*

The author believes that a ten-year moratorium should be placed on artificial inovulation research, the development of artificial wombs, attempts to clone small mammals and humans, cell fusion experiments, and recombinant DNA research. Weighing the pace of such research, its possible unpredictable results, and the potential societal problems, shocks, and danger which such research could bring, the author feels that no chances should be taken in these areas. He also favors controlling or slowing down the biological revolution in other areas as well, but feels that society may be able to more readily prepare for and adjust to the consequences of research in those other areas, by the time such developments arrive.

What should be done during the moratorium? The author feels that national and international, private, governmental, and scientific conferences and forums should be established to discuss such research, with far more opportunity for the cross-fertilization of ideas to take place than has usually been possible in the past. Perhaps there should actually be an International Biological Year just as there was an International Geophysical Year in the late 1960's. The leaders of every discipline and branch of learning should have an opportunity to come together, learn of, discuss, and plan for the implications and possible consequences of not only the parts of the biological revolution affected by the moratorium, but the biological revolution as a whole.

It should be emphasized that there are several precedents for such a moratorium. First of all, of course, there was the 18 month, self-imposed, voluntary moratorium on recombinant DNA research begun by scientists themselves in 1973. However, it could also be pointed out that the National Institutes of Health banned experimentation on live fetuses for almost two years beginning in April of 1973. And, on May 1 of 1972, the *Journal of the American Medical Association* called for a moratorium on experiments aimed at the development of "test tube babies." Lastly, the action of Congress and the people in banning further work on the SST in 1972 should be evidence that there may actually be more sympathy for such suggestions in the country than many think.

It's probably now time to make a second major point about the biological revolution. The point might be summarized by simply saying that "knowing is not enough." However, elaborating, it should be emphasized that *acknowl-*

edging the extremely serious societal problems which the biological revolution could create, leaves one with a very complex decision which must then be made —what to do about it. At first glance, it appears there is no answer. It would appear that it will be unreasonable to simply ignore, accept, or even endorse the biological revolution, difficult to control it or slow it down, and almost impossible to stop or prohibit it. The most realistic of the three options may be the second, an attempt to control its pace, which would essentially be a compromise between the other two, more extreme options. A 10 year moratorium on certain parts of the biological revolution would not seem to be an unreasonable price to pay for lessening future shock, at a minimum, and "getting a handle on" developments which could conceivably destroy us, at a maximum!

THE ROADBLOCKS TO POSSIBLE ACTION

Most people today have very little knowledge of what is happening in the biological revolution. And the majority of those remaining are largely ignoring, accepting, or endorsing it. There is thus very little sentiment, and almost zero effort, being directed toward attempting to prohibit or stop such developments. In fact, there are so many roadblocks looming in the way that any attempts to even control or slow its pace will be extremely difficult. The roadblocks are extremely formidible, for they are unfortunately the same obstacles that now block action on any major problem in society today.

Others may disagree, but the author feels that a partial list of such roadblocks in the United States would include the following: (1) an increasing immobilization of societal action, caused by experts who increasingly disagree on action alternatives; (2) a traditional gravitation toward optimism and over-optimism when confronted with conflicting action alternatives; (3) a traditional "back door" feeling that the government, science, or God will "save the day" before things could actually get out of hand; (4) increasing boredom, apathy, and psychological numbness after two decades of seemingly endless problems; (5) an increasing feeling of powerlessness; (6) a loss of wonder, mystery, and awe, producing an inability to be amazed or shocked; (7) decreasing reverence for life and increasing irresponsibility toward God and nature; (8) a traditional feeling of being nature's "superior animal"; and (9) increasing softness, selfishness, greed, and irresponsibility, being created by increasing materialism. Unfortunately, the list could be expanded even further. In fact, the biggest societal roadblock of all (and the least recognized) hasn't even been mentioned yet—a lack of societal wisdom. It's a roadblock which looms so large today that it demands more than just passing mention.

WHERE IS THE WISDOM?

Before the advent of modern science and technology, the words "knowledge" and "wisdom" were largely synonymous, and were often used interchangeably. Both words could have been broadly defined as being "the sum of what man knew," with the emphasis in that "sum" usually more on human nature than nature itself. Thus, for example, when the Hebrew prophet Hosea wrote in the 8th century B. C., "My people are destroyed for lack of knowledge," he could have just as easily used the word "wisdom," with no real loss in meaning. Hosea was speaking of inner discernment, insight, judgment, values, character, etc.—the kind of inner, spiritual "knowledge" that can change the human heart.

Since the advent of modern science and technology, however, our accumulative understanding of nature, which might be called our "outer" understanding, has increasingly outstripped our accumulative understanding of human nature, or "inner" understanding. One of the results of this widening gap has been a corresponding divergence in the meaning of the words "knowledge" and "wisdom," and a loss of their former synonymity. On the one hand, as our "inner" understanding has lagged, the word "wisdom," still associated with such understanding, has fallen into increasing disuse. But, on the other hand, the word "knowledge" has become associated with an increasing quest for "outer" understanding—a quest which now seems to largely define "the measure of man." "Knowledge" has thus increasingly come to mean an acquaintance with or understanding of a highly specialized science, art, or technique—with the extent to which one can master and utilize associated facts, figures, and equations being a measure of that knowledge.

"*Knowledge and wisdom are no longer synonymous.*" If one accepts that proposition as being at least partially representative of reality today, there are four additional, inferential statements which can be drawn from it that are at least debatable.

1) IT'S "KNOWLEDGE," HAPPINESS, CONTENTMENT, AND TO SOME DEGREE, EDUCATION, THAT ARE LARGELY THOUGHT TO BE SYNONYMOUS TODAY.

Those people in society today with the greatest "accumulation of knowledge" (as in the professions, for example) more often than not also have the greatest accumulation of material wealth as well. And thus, since material wealth, happiness, and contentment have always been thought to be synonymous by most, "knowledge," happiness, and contentment are now also usually thought to be synonymous! The route to happiness and contentment is usually thought to be "the pursuit of knowledge," with the vehicle for that pursuit being education. The idealistic goal of education is thus usually stated to be the ever-widening and more efficient dissemination of "knowledge."

2) THERE IS A CORRELATION BETWEEN "KNOWLEDGE," HAPPINESS, AND CONTENTMENT, BUT THERE IS A MISSING DIMENSION IN THE CORRELATION.

Although "knowledge" has supposedly quadrupled since about 1950, for the vast majority of people, happiness and contentment apparently have not!

There is documentable evidence that there have been major increases (usually out-stripping population growth) in all of the "visible," "negative" social indicators (e.g., crime, drug use, illiteracy, etc.) in the last generation or so. Less documentable evidence indicates there have also been extensive "negative" shifts in many "less visible" societal attitudes and codes of conduct such as self interest, greed, lack of concern for others, pessimism, etc. *Increasing* happiness and contentment? Many are now openly saying something has gone seriously wrong in America!

3) THE WIDENING GAP BETWEEN "KNOWLEDGE" AND "WISDOM" MAY IN LARGE MEASURE BE RESPONSIBLE FOR THE "SOMETHING" NOW APPARENTLY WRONG IN SOCIETY.

It would be highly controversial to contend that we suffer literally from too much "knowledge" today. And it would be degrading to think that we are literally no wiser now than man was in Hosea's day. It might be more palatable to simply assert that the *gap* between our "inner" and "outer" understanding has grown intolerably wide, and that, thus, *relatively speaking* we currently suffer from too much "knowledge" and not enough "wisdom." Our ethics haven't kept up with our physics! Relatively speaking, Albert Schweitzer was thus probably right when he once said, "The more man becomes superman, the more inhuman he becomes."

4) "WISDOM" IS THE MISSING DIMENSION IN THE AFOREMENTIONED CORRELATION—SOCIETY TODAY NEEDS A MASSIVE INJECTION OF "WISDOM"!

Values, purpose, goals, morality, ethics, inner discernment, character, etc. might be termed the vitamins of life. Using that analogy, the major and perhaps ultimate disease of our age could be diagnosed as being societal vitamin deficiency! Some have also called it "value-illness." The prognosis, if the disease continues? Increasing moral amnesia! An increasing loss of moral perspective; an increasingly foggy trip through history, with no clearly defined purpose, goals, or destination; an increasingly hollow, empty, "do-*your*-thing" society; and a gradual lapse into anarchy! The prescription? A massive societal injection of "wisdom"! Unfortunately, however, most of us don't even seem to realize we're sick!

T. S. Eliot eloquently summarized both our sickness and its consequences many years ago when he wrote,

> The Eagle soars in the summit of Heaven,
> The Hunter with his dogs pursues his circuit.
> O perpetual revolution of configured stars,
> O perpetual recurrence of determined seasons,
> O world of spring and autumn, birth and dying!
> The endless cycle of idea and action,
> Endless invention, endless experiment,

Brings knowledge of motion, but not of stillness;
Knowledge of speech, but not of silence;
Knowledge of words, and ignorance of the Word,
All our knowledge brings us nearer to our ignorance,
All our ignorance brings us nearer to death,
But nearness to death no nearer to God.
Where is the Life we have lost in living?
Where is the wisdom we have lost in knowledge?
Where is the knowledge we have lost in information?
The cycles of Heaven in twenty centuries
Bring us farther from God and nearer to the Dust.[4]

THE NEEDED REVOLUTION

It is wisdom that is an individual's and a nation's rudder and sets a proper course. When either an individual or a society no longer seeks and values wisdom, the course is lost and the ship begins to drift. Has this already happened to us? Was Garner Ted Armstrong right when he recently wrote the following in *Plain Truth* magazine?

> Most individuals today find themselves adrift. They go from one accident, one chance encounter, and one moment to the next—always trying to solve the problem after it has occurred, never trying to avoid it before it occurs.
>
> And so, we as a nation seem to be morally, spiritually, and emotionally adrift in life, sitting on the dock of the bay watching the tide go in, not knowing *what* we are, *why* we are here, and *where* we are going.
>
> Today, the United States is in a *crisis of the spirit.*[5]

It should be obvious that the biological revolution is just one of the many problems sweeping us along today in an uncharted sea of confusion. If we are going to solve any of those problems, we're obviously going to have to get our rudder back somehow. The idea is not new. H. G. Wells once warned that man should "get wise as soon as possible." T. S. Eliot said events had gotten "soul-size." He warned they can no longer be solved only on a technological level, but demand a realignment of the soul, the heart, and the deepest desires and motivations of human beings. In 1964, General Douglas MacArthur said man's only hope was an "improvement of human character." In 1975, Garner Ted Armstrong said it all even more pointedly when he said we need a revolution—"not a revolution of guns, but a revolution of the human spirit."[6]

An increasing number of futurists today are now saying basically the same thing, although they use somewhat more sophisticated terminology. They are calling for a "paradigm shift," a change in consciousness that will produce a new overall outlook on life. In his 1976 book, *The Global Mind,* Lewis J. Perelman uses seven characteristics to summarize both the old ("hardworld") paradigm under which he believes man is largely operating, and the new ("softworld") paradigm under which he says man must soon operate if he is to survive. The two paradigms are contrasted below. The terminology used should be largely self-explanatory.

Today's "Hardworld Paradigm"	*The Needed "Softworld Paradigm"*
1. Narrowmindedness	1. Holistic, Systems Thinking
2. Simplicity	2. Sophistication
3. Dogmatism	3. Flexibility
4. Scientific, Technological Fatalism	4. Scientific, Technological Skepticism
5. Short-Range Perspective	5. Long-Range Perspective
6. Irresponsibility	6. Responsibility
7. Violence	7. Peacefulness, Gentleness

THE "NEW" PERSON

A paradigm shift of the type just outlined would indeed require (and would constitute) a "revolution of the spirit." However, if it could be pulled off, the product would be a "new" person. That new person would be less self-centered, less orientated toward seeking power, and more concerned about nature, mankind, humanity, and future generations than are most people today. He would be less concerned about I, Me, and Thou, and more concerned about WE! He would be less afflicted with dogma and -isms, more interdisciplinary, holistic, and flexible in his thinking, and would increasingly think in terms of systems analysis. His increased sense of caution, self-restraint, and self-discipline, and his longer-term perspective would no longer allow him to believe in progress simply for the sake of progress and knowledge simply for the sake of knowledge. He would support rather than reject technology assessment and the regulation and self-regulation of science and technology. He would know when to forego as well as seize opportunities, and "when in doubt," he would "GO SLOW."

In summary, the "new" person would have what Dennis Gabor has called an improved "ethical quotient."[7] On Gabor's scale, a 70 minus is a "brutish, malicious, cruel, habitual criminal type of person"; 70 to 100, a person with a decreasing inclination to envy, hatred, dishonesty, low standards, lying, selfishness, and "kicks"; 100 to 110, a person "responsible and reliable in the right environment, but prone to accept the standards of the majority"; 110 to 120, "balanced attitudes between ego and social environment"; 120 to 130, "dedicated to socially useful works but ego not suppressed"; 130 plus, a person "dedicated to good works and the service of others to the point of self-effacement or even self-sacrifice." Today's "hardworld" paradigm probably confines most people to an ethical quotient range of 80 to 120. However, the new "softworld" paradigm would allow people to operate in the 120 and up range!

The whole subject of paradigm shift and how it might specifically apply to the scientist, layman, educator, and religious person should obviously be explored in great depth at this point. Unfortunately, however, it will be possible to make only a few more random comments in this respect about each of these types of people.

THE NEW SCIENTIST

In addition to exercising increased humility, social concern, caution, responsibility, foresight, and wisdom in his work, the new scientist would be willing to step down from his role of being society's new priest, and mingle with his parishioners! First of all, realizing that there are more pathways to knowledge and truth than just his own, he would no longer "look down his nose" at the opinions of theologians, philosophers, humanists, laymen, and even other scientists. In fact, he would not only welcome their advice, but perhaps even seek it, rather than wait for it to filter "up" to him. He would realize that although he's an expert on nature, others are experts on human nature. And thus, he would work harder in the future to develop the hazy interface that now exists with such people, than he has in the past. He would work harder at becoming the humanized scientist that society has needed for so long.

Secondly, the new scientist would do everything he could to increase public understanding of his work through science communication courses, adult education classes, public speaking, service on public commissions, institutes, forums, panels, articles, books, and science fiction written for the layman, and by making his own courses as relevant, society-orientated, and future-orientated as possible. And, while doing such things, he would also begin to devote as much attention as possible to such ultimate questions as "Who am I?," "Where am I headed?," "What should the world become?," etc.

Thirdly, the new scientist would at least consider a possible role for himself in politics, as a powerful way to possibly both inform and change society. And, lastly, those new scientists who do not have the time, energy, and talents to do as many of the above things as they might like, would find room under the umbrella of science for those who do—something not always done today!

THE NEW EDUCATOR

The "new" educator would realize that he also bears a major responsibility for informing the public about the progress of science and technology. But he would also realize that education must bear at least some of the responsibility for the "injection of wisdom" needed to treat society's "value deficiency." In making the following comments in this regard, the author will use the personal pronoun we since he considers himself to be an educator first and a scientist second.

Montaigne (and others all through history) maintained that "the object of education is to make, not a scholar, but a man." However, education today is all too often an ever-narrowing tunnel, filled with a blinding smokescreen of teaching "innovation." Graduates all too often stagger out, afflicted with tunnel vision—they know more and more about less and less! Highly-specialized, they know technically *how* to change the world, but usually haven't the foggiest notion of what to change it into! In fact, all too often they don't even seem to *care* about the answers to the big questions of human existence.

How can we put wisdom—the missing dimension in education today—back in? The long-range answer, of course, is that we must get back to the earlier, historical definition of education, and provide students with a more liberal education. The new educator must increasingly turn out a new product —a new breed of Renaissance man! In the 1960's, we sold the idea of "job training" to those students who wanted "relevance." In the 1970's and 1980's, the new educator must resell them on relevance!

The long-range goal will admittedly be extremely difficult to achieve from within the educational system, for reasons that should be obvious. But there is one thing that we can do immediately as we attempt it. We can *all* turn to the best of our abilities from being only or largely "imparters of knowledge" to also being "professors of wisdom." We can turn from a rhetoric of non-commitment to a rhetoric of commitment. We can start discussing *why* as well as *how*, and pose and discuss ultimate as well as more temporal questions. We can stop hiding behind both determinism and relativism and, as professors begin to profess, start risking judgments and taking stands. We can turn from simply intellectual and technological entropy to attempting to recrystallize the positive values of life. We can attempt to no longer leave students in an intellectual and moral vacuum, but make them positive disciples as well as permissive skeptics. We can attempt to better answer the question of "Who am I?" and "Where am I going?" as we continue to tackle the question of "What am I?" In short, we can attempt to make them "men" as well as scholars!

There probably isn't a course currently on the books that couldn't be modified in this way to at least some degree, without short-circuiting "the pursuit of knowledge." We can liberalize education immediately, without scrapping specialization! How would we know we have succeeded? In 1974, the Rockefeller Brothers Fund (RBF) sponsored a study at the Western Interstate Commission for Higher Education (WICHE) whose purpose was to explore such questions.[8] According to the report developed (titled "Growth and Education"), an "adequate" education should be (1) Multi-level (that is, should cultivate not only ordinary learning, but consciousness change as well; (2) Interdisciplinary; (3) Problem-centered; (4) Future-orientated; (5) Global; (6) Humanistic. These terms all obviously require further extensive explanation, but their primary or skeletal meanings should be obvious. The report went on to state that, in the author's opinion, there were curricula in 1974 which satisfied a few of the characteristics, but no curriculum in use which demonstrated all.

THE NEW LAYMAN

The "new" layman realizes, as Buckminster Fuller once said, that "Spaceship Earth didn't come with an instruction book." And he realizes that he is quite a few pages behind those who are attempting to write one. He thus realizes that he has some homework to do, but also that there aren't any experts on the future anymore on whom he can depend. And so he begins to formulate his own opinions on an instruction book. But while doing his own homework, he realizes that writing, math, and science skills are declining among the young, while "functional illiteracy" is on the increase. And thus he does what he can to uplift such skills in both his own children and others, for he knows that the future can only be molded by literate, informed people.

The new layman realizes that problems today are too non-technical to be decided only by technical people. And thus, once he formulates some opinions as to how the future should unfold, he begins to work with those groups or organizations in society through which he can inform and change society.

THE NEW RELIGIOUS PERSON

The "new" religious person realizes that there are no definitive mandates in the Bible or other religious writings, or in any religious tradition and heritage, for many of the questions and problems now facing society. That is particularly true in the case of the biological revolution. Where, for example, does one obtain historical advice for questions concerning artificial inovulation, cloning, transplants, genetic engineering, etc.? Men and women in other ages not only didn't discuss such developments, they didn't even dream of them!

About two decades or so ago, H. Richard Niebuhr called the church into responsibility when he said, "The church cannot be responsible to God for men without becoming responsible for their societies. . . . If the individual sheep is to be protected, the flock must be guarded." The new religious person accepts those words and attempts to use his knowledge to be a significant force for social progress and reform. He knows the task will not be easy if he decides to work through organized religion, and that there is a tremendous built-in psychological inertia in religion today when it comes to action directed toward science. In view of its past record with science, it will be difficult for the church to advocate control or prohibition of all or part of the biological revolution. By doing so the church would be once again looked upon as being out of its realm, and its advice, in a sense, would be considered as coming from the "loser" of many battles with science in the past.

The church is a sleeping giant! Its most common sin today is silence. And when it does speak, it all too often mumbles—usually after the fact. But it can no longer afford the luxury of attempting to speak to a problem after it arrives, and it can no longer afford to mumble. How long will society continue to look for guidance to a church that remains silent or mumbles on the key issues of our age? If the church really does have a responsibility to attempt to "protect the flock," shouldn't it also attempt to do so before as well as after the wolf arrives? The new religious person realizes he must attempt to wake the sleeping giant!

> Nothing, therefore, is more necessary than to arouse responsible Christians from their lethargy and slumber into the infinite dangers and infinite possibilities of the moment.
> —Charles Malik

> It is time that the church stop being like a reluctant little child, always needing to be dragged into the present.
> —Joseph T. McGuchen,
> Catholic Archbishop,
> San Francisco

> Christian moralists have a habit of arriving late on the scene. We come in as undertakers rather than as preventive medical officers.
> —Lord Soper, English Methodist spokesman

> How often the church has been an echo rather than a voice, a taillight . . . rather than a headlight guiding men progressively and decisively to higher levels of understanding.
> —Martin Luther King, Jr.

Is the church willing to leave its defensive fortifications? Is the church willing to speak out with regard to the future, even if it means challenging science again? Is the church willing to climb back into the ring with science again, even at the risk of being bloodied again? Is the church willing to make the first move in this respect? The new religious person believes all of these things *must* be done, and he does his best to help bring them about.

Even if the church believes that the biological revolution should not, cannot, or will not be controlled, limited, or slowed down, doesn't it at least have the responsibility to prepare society for the biological discoveries that are possibly coming? As part of that informational role, does the church have a responsibility to inform people how to make decisions in this area—how to exercise "the Christian responsibility" we so often talk about? In other words, assuming society is going to play God, should the church attempt to help society do it as responsibly as possible? As theologian Edwin Dahlberg said, "Our day is full of bewildering scientific discoveries. The chief responsibility of the world's religious thinkers is to interpret these discoveries in such a way as to give meaning and direction to life."[9]

At a minimum, doesn't the church at least have a responsibility to attempt to lessen "future shock" by providing opportunities for intellectual confrontation and serving as an arena for debate on the biological revolution? Or isn't the church interested in the possible prospect of Bio-Babel?

Lastly, the new religious person and organized religion would do mankind a tremendous service if they could help us all rediscover sin. As John M. Jensen said, writing in the August 5, 1975 issue of the *Lutheran Standard,* "Do not be fooled. The serpent is sneaking about everywhere—in our government, in our society, in our business world, in our social life, and also in our churches. The serpent's bite is deadly. So beware of him, ever in the garden."

Is the serpent also "sneaking about" in science and in the biological revolution? If sin is merely the separation of man from God, nature, and other men, is there sin involved in our science and in the biological revolution?

If the new religious person and organized religion were to do their homework on the biological revolution, a bonus would soon make itself clear: the use of science to reinforce religious beliefs. It might actually be possible to even reverse the trend today wherein science is eroding such beliefs. It needn't be so! There is a place for God in science, once we're willing to admit that we're only reading the book of science, not writing it. At that point one begins to realize that all truth ends in mystery. Human knowledge is circumscribed by an unexplored domain. Science thus always raises more questions than it answers. Science can answer the questions of what, when, where, and how—but not *why!*

For example, *why* do carbon atoms have so many unusual properties which make them, and only them, just right to be the basis of life? *Why* is the formation of life inevitable if the conditions are right? *Why* does life seemingly transcend the laws of chemistry and physics? *Why* does the whole in science so often seem to exceed the sum of the parts? *Why* does Nature proceed automatically to go about her business even if we don't understand her ways? *Why* is it preferable to say, that rather than God as the architect, mindless, inanimate, dead matter organized itself, became animated or alive, and endowed itself with consciousness and thought? As Loren Eiseley said in his book *The Immense Journey,* ". . . And if 'dead' matter has reared up this curious landscape of fiddling crickets, song sparrows, and wondering men, it must be plain even to the most devoted materialist that the matter of which he speaks contains amazing, if not dreadful, powers, and may not impossibly be but one mask of many worn by the great face behind."[10]

The previous list of questions could be greatly expanded, and a whole series of specific facts could be pulled from the biological revolution which could only increase wonder, mystery, and awe! For example, as one looks at such things as a synapse undergoing countless repetitions within a millionth of a second, at a "beating" mitochondria energy factory, at a cell as a miniature chemical factory, at a chromosome whose chemical information if printed out would fill 2000 volumes of about 1000 pages each, and at the amazing, even awesome complexity and yet simplicity of DNA, RNA, the protein "alphabet," the triplet or genetic code, etc., can we really deny the existence and operation of a thought system greater than our own? Could all of this really be a blind, random, cosmic accident? Or is one forced to say that there is repeatable order present in such things? Does such order imply an ordainer? There's a whole new branch of theology hidden (but not too well!) in the biological revolution—bio-theology!

There is room for God in nature, evolution, and science in general. In fact, if you think strongly enough, science almost forces you to believe in God. When organized religion fully realizes that science can be used to reinforce religious belief, and reach the young people of our age in a new way, a new tool of almost unfathomable power will be unleashed. And society will be the better for it. As Sydney J. Harris said recently in his syndicated newspaper column, "In the crisis of our times, the I can save itself only by reaching out to the Thou and saying 'We.' "

WHAT TO DO UNTIL THE REVOLUTION ARRIVES

It should be obvious that if we had the revolution, the societal injection of wisdom, and the paradigm shift (and the "new" scientist, layman, educator, and religious person which they would produce) just described, the biological revolution, and, in fact, all of the other problems of our age, would automatically take care of themselves. Is such a revolution too idealistically "blue sky" to ever hope for? The answer is—not necessarily!

In his 1962 book *The Structure of Scientific Revolutions,* Thomas S. Kuhn argues that science advances through revolutionary paradigm shifts. He says an accepted paradigm operates unquestioned and unopposed for a certain period of time, suppressing "fundamental novelties and change." But when the rules begin to fail, questions arise, and crises even develop for "normal science," at least some scientists begin to feel uneasy and begin to feel the need for paradigm change. A new, alternative paradigm is developed, which then goes through a period of debate, controversy, and even ridicule. But if a few switches of allegiance occur, they usually catalyze others, and a period of transition arises. It may be greatly reinforced by a "gestalt switch"—"overnight," "transformed," "see the light" types of a switch. Sooner or later the new paradigm is adopted, and the old paradigm discarded. Kuhn argues that paradigm switches are *not* incremental, step by step processes forced by logic and thought. Once dissatisfaction, uneasiness, and the desire for a new paradigm surface, the transition becomes a psychologically mushrooming process. He argues that societal paradigm shifts operate in the same way. If he's right about all of this, it's good news, for if a paradigm shift can be started, history may complete it! With history on your side, the problem of a radical paradigm shift may not be so formidable. Of course, the problem still remains of how to get it started sufficiently so it can begin to roll on its own.

And thus the first thing that the reader can do about the biological revolution is to prepare to do something. Do your "homework," decide where you are on the scientism-anti-science spectrum, and examine the paradigm under which you operate. Try to bring about a paradigm shift in yourself before you attempt to do the same in others. Do what you can to be the "catalyst" Thomas Kuhn describes above in bringing about the type of paradigm shift which was outlined earlier. And then take a stand on the general and specific action alternatives which have been outlined earlier in this chapter. Remember, as Dante once said, that, "Not to decide is to decide," and "The choicest spots in Hell are reserved for those who in times of crisis do nothing"! Be alert to the fact that, although options two and three can still be undertaken, they will narrow and even close as time goes on, for the biological revolution will gradually develop its own dynamics and take matters out of your hands. After making your choice, do what you can to advance it. Attempt to change your little corner of the world first, but also branch out if possible to work individually or collectively on a larger scale.

It is at this point that you will encounter three major roadblocks within yourself. The first is the feeling of lonely powerlessness that is so widespread today—the feeling that you are alone in attempting to do something that won't matter anyway. It doesn't always help either, to realize that, although you are only one person, you have a responsibility to do what you can as that one person. It might be helpful to remember at this point that for one of the few times in history you have an opportunity to speak to, mold, and perhaps even change history before it arrives. You have a chance to make a decision and act on it while there is still a maximum number of alternatives left from which to choose. You have a chance to be a "headlight" rather than a "taillight," as Martin Luther King once said. And, you should also remember that individuals have changed history, often without even intending or planning to do so. Rosa Parks was such a person. She said one word on the Cleveland Street Bus in Montgomery, Alabama on December 1, 1955, which helped change the world. She said the word when the bus driver asked her to get up and give her seat to a white man. The word was "no!" It put her in jail, but it also caused demonstrations which helped prompt the Supreme Court to act on desegregation. You may not be able to change the world to the extent Rosa Parks did—but then again you might!

Every revolution was first a thought in one man's mind. . . . Every reform was once a private opinion. —Emerson

Not armies, not nations, have advanced the race; but here and there, in the course of ages, an individual has stood up and cast his shadow over the world.
—E. H. Chapin

A pebble can affect the entire ocean. —Pascal

A man in the right, with God on his side, is in the majority though he be one.
—H. W. Beecher

The second roadblock that you will encounter within yourself will be a feeling that you will be criticized, penalized, or even hurt in your public, private, or personal life if you attempt to advance this revolution. You will be! All revolutionaries get hurt at some point in the revolution and you will be no different. Don't begin the revolution unless you understand that! But you might ask yourself whether you could dedicate your life to a more worthwhile goal than attempting to mold history and perhaps save mankind and future generations. What are your goals today?

He who has a firm will molds the world to himself.

Daring ideas are like chessmen moved forward; they may be beaten, but they may start a winning game. —Goethe

Lastly, don't be afraid of making a 20th century decision. The possibility of making a mistake and doing the wrong thing is simply part of being a 20th century man. A good guideline from which to operate in the 20th century can be found in the following poem based on excerpts from *Ethics* written by Dietrich Bonhoeffer. The poem was written by members of the Ecumenical Institute in Chicago.

Responsibility

Observe and judge the given facts.
Weigh up the values, decide, and act.
You're alone, completely free,
Leave the judgment to history.

To no principle, no law,
To no authority can you withdraw.
You decide it all alone,
Right from right and wrong from wrong.

Obligation is the call,
To God and neighbor surrender all.
The free venture is the deed.
Rendered up to meet the need.

> Free men live in responsibility.
> Duty bound and free in relativity.
> Free men live in responsibility,
> Whoever they may be,
> Their deeds are history.

Dietrich Bonhoeffer was a theologian living in Germany during Adolph Hitler's rise to power. Bonhoeffer had a difficult 20th century decision to make —whether to remain committed to a philosophy opposed to killing or join a plot dedicated to the assassination of Hitler. Bonhoeffer decided on the latter course of action, the plotters were caught, and Bonhoeffer was one of those hung!

The author has now revealed his "game plan" for dealing with the biological revolution. This chapter might thus be ended with two fables that have something to say in support of that plan.

First of all, there is an old tale which says that monkeys can be trapped through the use of a small-mouthed jug containing a large nut. The monkey sees the nut in the jug and reaches his paw in to take it. However, in grasping the nut, the monkey's paw becomes too large to be withdrawn from the jug. He could escape by simply releasing the nut. But he refuses to let go and thus is trapped.

The second fable is a modern fable.

> This little pig built a spaceship,
> This little pig paid the bill;
> This little pig made isotopes,
> This little pig ate a pill;
> And this little pig did nothing at all,
> But he's a little pig still.
> —Frederick Winsor, *The Space Child's Mother Goose*

CONCLUDING COMMENTS

> There is no assurance that science may not destroy man. The scientific exploration in which we are now engaged makes the Tower of Babel look like an experiment with building blocks. The consequences could be correspondingly terrible.[1]
>
> —Edwin Dahlberg, theologian

The outlines of another historical tower are slowly but inexorably coming into focus. The base is already rather sharply defined. It is comprised of an extremely complex set of contemporary and future physical and psychological problems, a lack of historical perspective, a lack of societal wisdom, and a "cloudy crystal ball." But that base, however shaky it may be, must now support the added weight of the biological revolution—perhaps the greatest scientific revolution of all time. Unfortunately, the exact future form, dimensions, and weight of this added structure are currently unknown.

Is the biological revolution an "epic opportunity" or simply "a potential too large to grasp"? Will it bring to fruition a goal which evolution may have been nurturing for 3.5 billion years—for mankind and nature to rise above blind, random chance—or will it thwart God's plans for man and the natural scheme of things so severely that mankind could face the danger of extinction? Is terrestrial man rational, ready, and wise enough to intelligently use the cosmic knowledge and power which he has now grasped? Is mankind building the most impressive tower of knowledge ever—a tower whose top will reach the heavens? Or, is he building a Bio-Babel which will one day lie in ruins at his feet?

> The success or failure with which we "play God" in the next few years will determine whether these are the first few moments in mankind's greatest and most exciting hour or the last few seconds in his ultimate tragedy.[2]
>
> —Leroy Augenstein, physicist and theologian

Thus the slow-working accidents of nature, which by the very patience of their small increments, large numbers, and gradual decisions may well cease to be "accident" in outcome, are to be replaced by the fast-working accidents of man's hasty and biased decisions, not exposed to the long test of the ages. His uncertain ideas are to set the goals of generations, with a certainty borrowed from the presumptive certainty of the means. The latter presumption is doubtful enough, but this doubtfulness becomes secondary to the prime question that arises when man indeed undertakes to "make himself": in what image of his own devising shall he do so, even granted that he can be sure of the means? In fact, of course, he can be sure of neither, not of the end, nor of the means, once he enters the realm where he plays with the roots of life. Of one thing only can he be sure: of his power to move the foundations and to cause incalculable and irreversible consequences. Never was so much power coupled with so little guidance for its use.[3]

—Hans Jonas, philosopher

We are witnessing the erosion, perhaps the final erosion, of the idea of man as something splendid or divine, and its replacement with a view that sees man, no less than nature, as simply more raw material for manipulation and homogenization. Hence, our peculiar moral crisis. We are in turbulent seas without a landmark precisely because we adhere more and more to a view of nature and of man which both gives us enormous power and, at the same time, denies all possibility of standards to guide its use. Though well-equipped, we know not who we are nor where we are going. We are left to the accidents of our hasty, biased, and ephemeral judgments.

Let us not fail to note a painful irony: our conquest of nature has made us the slaves of blind chance. We triumph over nature's unpredictabilities only to subject ourselves to the still greater unpredictability of our capricious wills and our fickle opinions. That we have a method is no proof against our madness. Thus, engineering the engineer as well as the engine, we race our train, we know not where.[4]

—Leon R. Kass, biologist

The future doesn't descend upon us on some prophetic day like the Messiah. . . . It grows out of forces which are now turbulently in motion. . . . Whatever warnings we're going to get we already have.[5]

—Theodore Levitt, educator

At some particular state in scientific development will the good Lord, with a flowing white beard, arrive on earth with his chain of keys and say to humanity, just like they do at the Art Gallery at 5:00 . . . "Gentlemen, it's closing time."[6]

—Jules de Goncourt

POSTSCRIPT

Most books are at least somewhat out-of-date by the time they appear and this one is unfortunately no exception. The research and developments discussed previously occurred prior to the end of April of 1977. However, it might be added that in just the next six months:

A new psychochemical treatment was announced for treating and counteracting senility.[1]

Major progress was announced in the development of a new gene mapping process. The process, which has already been used to largely map a gene that directs the production of a human growth hormone, is being hailed as possibly opening up the path to studying and mapping almost any human gene.[2]

The first successful transplantation of an insulin-producing gene into *E. coli* (by recombinant DNA techniques) was announced, promising the production of insulin by bacteria "right around the corner."[3]

The concern and controversy surrounding recombinant DNA research sharpened and reached new heights.[4]

The San Francisco State University's Science-Humanities Convergence Program announced it was considering the "shocking" idea that not only may aggression, altruism, and other human behavior be transmitted genetically, but homosexuality may be as well!

Dr. Christiaan Barnard transplanted the first baboon heart into a human being.

The first testicle transplant (between twins) was announced.[5]

New clues were reported found as to how genes switch on and off during development.[6]

One of the enkethalins previously found to be a pain killer in the brain has also now been reported to improve learning abilities.[7]

A new drug (bromocryptine) was reported, which not only corrected sexual disorders in certain patients with hormonal deficiencies, but also increased sexual libido and potency in the same patients.[8]

Further progress was reported in developing artificial hearts, kidneys, and limbs.[9]

Startling evidence was reported that the gene product in higher organisms may not be represented simply by a linear, uninterrupted sequence of nucleotides. In other words, the code of life may be far more complex than previously thought.[10]

Insulin gene researchers at the University of California were accused of the first violation of the National Institutes of Health guidelines on reconbinant DNA research.[11]

The biological revolution marches on!

1. *SN,* Volume 111, May 7, 1977, p. 292.
2. *SN,* Volume 111, May 7, 1977, p. 295; *CEN,* May 9, 1977, p. 16.
3. *Newsweek,* June 6, 1977, p. 74.
4. *SN,* Volume 111, May 7, 1977, p. 293; *CEN,* May 30, 1977, p. 26.
5. *Newsweek,* June 6, 1977, p. 39.
6. *SN*, Volume 112, July 23, 1977, p. 54.
7. *SN*, Volume 112, July 23, 1977, p. 39.
8. *SN*, Volume 112, August 13, 1977, p. 105.
9. *SN*, Volume 112, September 3, 1977, p. 154.
10. *SN*, Volume 112, October 1, 1977, p. 214.
11. *SN*, Volume 112, October 1, 1977, p. 212.

End

The judge sat stern; He had to find,
Whose fault it was—who changed mankind.
The evidence was clearly shown,
'Twas man himself who must atone.
He opened up his "Book of Time,"
And noted details of the crime.

"I breathed in life to every man,
I gave you love, you turned and ran.
And as a child must disobey,
You too did wrong and had your way.
You ventured in the dark unknown,
You opened doors were mine alone.
Too late you found mistakes were made,
Now you've a debt that must be paid."

He sadly uttered one last breath,
"The verdict chosen you is death."
The judge then closed his "Book of Time";
The bells began a mournful chime.
And as he walked the distant hill,
The darkness fell and all lay still.

—Janice L. Utke

SUGGESTED FURTHER READING

The following list is arranged chronologically over the last several decades. The list contains suggestions for further reading by the author, but is only a very brief introduction to the literature available on the biological revolution and related topics.

Books Directly Related To The Biological Revolution:

1. *The Biological Time Bomb,* Gordon Rattray Taylor, The World Publishing Co., 1968.
2. *The Prometheus Project,* Gerald Feinberg, Doubleday, 1968.
3. *The Second Genesis: The Coming Control of Life,* Albert Rosenfield, Vintage, 1969.
4. *Come Let Us Play God,* Leroy Augenstein, Harper & Row, 1969.
5. *Fabricated Man,* Paul Ramsey, Yale University Press, 1970.
6. *Biology in the World of the Future,* Hal Hellman, M. Evans Co., 1971.
7. *The New Genetics and the Future of Man,* edited by Michael Hamilton, William B. Eerdmans Co., 1972.
8. *The New World in the Morning: The Biopsychological Revolution,* David P. Young, Westminster, 1972.
9. *Genetic Fix,* Amitai Etzioni, Macmillan, 1973.
10. *Ethical Issues in Human Genetics,* edited by Bruce Hilton and Daniel Callahan, Plenum, 1973.
11. *Pre-meditated Man: Bioethics and the Control of Future Human Life,* Richard M. Restak, Viking, 1975.
12. *Biomedical Ethics and the Law,* edited by James M. Humber and Robert F. Almeder, Plenum, 1976.
13. *Biohazard*, Michael Rogers, Knopf, 1977.
14. *Playing God*, June Goodfield, Random House, 1977.
15. *The Ultimate Experiment*, Nicholas Wade, Walker, 1977.

Books Indirectly Related To The Biological Revolution:

1. *Brave New World,* Aldous Huxley, Harper & Row, 1932.
2. *Science Is a Sacred Cow,* Anthony Standen, E. P. Dutton, 1950.

3. *Religion and the Modern Mind,* W. T. Stace, Lippincott, 1952.
4. *Brave New World Revisited,* Aldous Huxley, Harper & Row, 1958.
5. *The Limitations of Science,* J. W. N. Sullivan, Mentor Books, 1961.
6. *The Universe: Plan or Accident?,* Robert E. D. Clark, Muhlenberg Press, 1961.
7. **The Structure of Scientific Revolutions,* Thomas S. Kuhn, The University of Chicago Press, 1962.
8. *Profiles of the Future,* Arthur C. Clarke, Bantam Books, 1967.
9. *Crisis In Eden,* Frederick Elder, Abingdon Press, 1970.
10. *Future Shock,* Alvin Toffler, Random House, 1970.
11. **Steps to an Ecology of Mind,* Gregory Bateson, Ballantine, 1972.
12. *Science and Anti-Science,* Morris Goran, Ann Arbor Press, 1974.
13. *The Global Mind,* Lewis J. Perelman, Mason/Charter, 1976.

*Excellent discussions of paradigm shift.

NOTES

This book is written primarily for laymen and those with little or no technical and/or scientific training. Thus, rather than citing highly technical articles in often little-known and difficult-to-obtain journals, whenever possible the author has referred to more popular and easier-to-obtain sources of information that summarize the original research. The three sources most often referred to will be abbreviated as follows: *SN = Science News CEN = Chemical and Engineering News IR = Industrial Research*

Those interested in seeing the original journal articles can obtain a more technical bibliography by writing the author at:

Chemistry Department
University of Wisconsin-Oshkosh
Oshkosh, Wisconsin 54901

Preface

The quotation by Angelo J. Cerchione can be found in *Energy Crisis: Danger and Opportunity,* edited and written by Victor J. Yannacone, Jr. (West Publishing Co. 1974), p. 409.

Part I

Chapter 1

1. *CEN,* May 26, 1975, p. 17.
2. *CEN,* February 17, 1975, p. 23; *SN*, Volume 104, November 10, 1973, p. 296.
3. *SN,* Volume 109, February 21, 1976, p. 118; *SN,* Volume 110, August 28, 1976, p. 133.
4. *SN,* Volume 100, July 17, 1971, p. 37; Volume 103, February 10, 1973, p. 93.
5. *SN,* Volume 107, January 4, 1975, p. 4; *CEN,* January 10, 1977, p. 21.
6. *SN,* Volume 106, August 3, 1974, p. 85.
7. *CEN,* April 21, 1975, p. 41.
8. *SN,* Volume 109, June 19, 1976, p. 390.
9. *SN,* Volume 103, February 10, 1973, p. 93; Volume 105, May 1, 1974, p. 309.
10. *Newsweek,* April 1, 1974, p. 50.
11. *SN,* Volume 107, April 5, 1975, p. 226.

12. *CEN,* October 29, 1973, p. 19.
13. *CEN,* May 12, 1975, p. 16.
14. *SN,* Volume 105, January 12, 1974, p. 20.
15. *Nature,* Volume 242, December 14, 1973, p. 421.
16. *SN,* Volume 110, October 9, 1976, p. 231.
17. *SN,* Volume 108, December 6, 1975, p. 356.
18. *SN,* Volume 110, October 11, 1976, p. 7.
19. *SN,* Volume 94, December 14, 1968, p. 591; Volume 103, February 24, 1973, p. 124.
20. *CEN,* April 9, 1973, p. 11.
21. *SN,* Volume 103, June 30, 1973, p. 419.
22. *SN,* Volume 107, February 1, 1975, p. 73.
23. *Science Digest,* January, 1974, p. 28; *Newsweek,* July 29, 1974, p. 70.
24. "The Test Tube Baby Is Coming," *Look,* May 18, 1971, p. 83.
25. *SN,* Volume 104, September 15, 1973, p. 168.
26. *SN,* Volume 106, July 20, 1974, p. 37.
27. *SN,* Volume 100, September 25, 1971, p. 205.
28. *Science Digest,* February, 1966, p. 40; Lino LoBello, "Russia's Test Tube Baby," *Argosy,* April, 1970, p. 34.
29. *Newsweek,* September 2, 1974, p. 76.
30. *Chemistry* magazine, Volume 46, number 7, July-August, 1973, p. 10.
31. J.B. Gurdon, "Transplanted Nuclei and Cell Differentiation," *Scientific American,* December, 1968, p. 24.
32. *Chemistry,* Volume 42, Number 5, May, 1969, p. 20; *Scientific Research,* April 28, 1969, p. 17.
33. *SN,* Volume 109, March 27, 1976, p. 197.
34. Albert Rosenfield, "The New Man—What Will He Be Like?" Part 4, *Look,* October 1, 1965, p. 98.

Chapter 2

1. *SN,* Volume 106, November 16, 1974, p. 315.
2. *The National Enquirer,* October 26, 1975, p. 6.
3. *CEN,* July 8, 1974, p. 22.
4. *CEN,* April 5, 1971, p. 32; April 12, 1971, p. 68.
5. *CEN,* May 17, 1976, p. 24.
6. *CEN,* May 24, 1976, p. 12; May 17, 1976, p. 24.
7. *The National Enquirer,* December 16, 1973, p. 30.
8. *Newsweek,* November 2, 1970, p. 100; *CEN,* May 18, 1970, p. 30; *SN,* Volume 106, September 28, 1974, p. 202.
9. *Newsweek,* March 24, 1975, p. 91.
10. *The National Enquirer,* April 9, 1972, p. 18.
11. *SN,* Volume 107, April 26, 1975, p. 268.
12. *CEN,* February 4, 1974, p. 13.
13. *CEN,* April 28, 1975, p. 20.
14. *SN,* Volume 100, November 13, 1971, p. 322.
15. *SN,* Volume 101, April 1, 1972, p. 218; Volume 110, July 17, 1976, p. 42.

16. "Researchers Believe They Have Solved the Mystery of Life," *The Milwaukee Journal,* October 25, 1970, p. 1.
17. "Electrochemistry in Organisms," *Journal of Chemical Education,* Volume 52, number 2, February, 1975, p. 99.
18. *SN,* Volume 108, October 11, 1975, p. 234.
19. *SN,* Volume 111, March 5, 1977, p. 149.
20. *CEN,* May 25, 1970, p. 11; *Newsweek,* June 1, 1970, p. 63; *SN*, Volume 108, July 5, 1975, p. 13.
21. *The Birmingham Star,* August 15, 1972, p. 17.
22. *SN,* Volume 108, August 23 and 30, 1975, p. 121.
23. *SN,* Volume 104, December 15, 1973, p. 373.
24. *SN,* Volume 106, July 6, 1974, p. 5.
25. *SN,* Volume 105, March 2, 1974, p. 134.
26. *CEN,* December 9, 1974, p. 19.
27. *Chemistry* magazine, Volume 45, Number 4, April, 1972, p. 26; *SN,* Volume 100, October 23, 1971, p. 276.
28. *Reader's Digest,* December, 1975, p. 135; *SN,* Volume 109, June 19, 1976, p. 389.
29. *Newsweek,* January 12, 1976, p. 50.
30. *Newsweek,* June 17, 1974, p. 54.
31. *SN,* Volume 111, April 2, 1977, p. 217.
32. Herbert W. Boyer, Robert B. Helling, et al., "Replication and Transcription of Eukaryotic DNA in *E. coli,*" *Proceedings of the National Academy of Sciences,* May, 1974, p. 1743.
33. *SN,* Volume 105, June 1, 1974, p. 348.
34. *SN,* Volume 109, April 10, 1976, p. 231.
35. *CEN,* June 14, 1976, p. 21.
36. *CEN,* June 8, 1970, p. 9; *Newsweek,* June 15, 1970, p. 91.
37. *CEN,* September 17, 1973, p. 18; *SN,* September 1, 1973, p. 132; *CEN,* September 21, 1974, p. 80.
38. *CEN,* September 4, 1976, p. 149; September 20, 1976, p. 27; *Newsweek,* September 13, 1976, p. 72.
39. *SN,* Volume 101, February 5, 1972, p. 86.
40. *SN,* Volume 108, December 13, 1975, p. 372.
41. *CEN,* February 25, 1974, p. 19.
42. *SN,* Volume 105, January 19, 1974, p. 41.
43. *SN,* Volume 109, May 22, 1976, p. 330.
44. *IR,* February, 1975, p. 17; *CEN,* December 2, 1974, p. 16.
45. *CEN,* January 7, 1974, p. 22.
46. *Plain Truth* magazine, March, 1977, p. 39.
47. *SN,* Volume 106, November 30, 1974, p. 293.
48. *SN,* Volume 106, November 30, 1974, p. 349.
49. *CEN,* February 16, 1976, p. 32.
50. *CEN,* December 14, 1970, p. 15; *Bulletin of Atomic Scientists,* December, 1972, p. 20, published by the Educational Foundation for Nuclear Science.
51. *Newsweek,* July 1, 1974, p. 76; *SN,* Volume 105, June 22, 1974, p. 397.

52. *SN,* Volume 111, March 26, 1977, p. 197.
53. *CEN,* September 13, 1976, p. 20.
54. *SN,* Volume 102, August 26, 1972, p. 133.
55. *Science,* July 30, 1976, p. 85; *SN,* Volume 110, July 31, 1976, p. 70.
56. Paul Ramsey, *Fabricated Man* (Yale University Press, 1970), p. 77.

Chapter 3

1. *SN,* Volume 105, February 16, 1974, p. 106.
2. *Newsweek,* September 28, 1970, p. 88.
3. *SN,* Volume 106, July 23, 1972, p. 183.
4. *SN,* Volume 110, October 30, 1976, p. 283.
5. *SN,* Volume 108, November 22, 1975, p. 327.
6. *SN,* Volume 109, May 15, 1976, p. 309; Volume 109, June 26, 1976, p. 406; *CEN,* August 16, 1976, p. 19.
7. *Newsweek,* February 11, 1974, p. 48.
8. *SN,* Volume 109, January 31, 1976, p. 68.
9. *IR,* November 15, 1975, p. 74.
10. *Scientific Research,* September 30, 1968, p. 30; *Intellectual Digest,* September, 1971, p. 69.
11. *CEN,* October 26, 1970, p. 16; *SN,* Volume 98, October 24, 1970, p. 331.
12. *SN,* Volume 101, May 20, 1972, p. 334.
13. *IR,* June, 1972, p. 33.
14. *SN,* Volume 101, May 20, 1972, p. 334.
15. *SN,* Volume 104, October 6, 1973, p. 218.
16. *SN,* Volume 107, February 8, 1975, p. 83.
17. *Science Digest,* May, 1968, p. 7; *CEN,* November 3, 1969, p. 43.
18. *CEN,* January 11, 1971, p. 27; *SN,* Volume 103, March 24, 1973, p. 180; *CEN,* November 26, 1973, p. 180.
19. *Chemistry* magazine, Volume 46, Number 7, July-August, 1973, p. 21.
20. *CEN,* April 28, 1975, p. 22.
21. *SN,* Volume 109, March 23, 1976, p. 169.
22. *Look,* November 28, 1967, p. 99.
23. *Newsweek,* December 3, 1973, p. 79.
24. *SN,* Volume 104, October 20, 1973, p. 247.
25. *SN,* Volume 110, October 30, 1976, p. 277.
26. *SN,* Volume 105, April 20, 1974, p. 155; Volume 108, July 12, 1975, p. 22; Volume 108, October 18, 1975, p. 224; Volume 109, April 3, 1976, p. 219.
27. *SN,* Volume 111, March 19, 1977, p. 184.
28. *IR,* November 15, 1975, p. 16.
29. *SN,* Volume 111, January 1, 1977, p. 8.
30. *SN,* Volume 109, February 28, 1976, p. 133.
31. *IR,* July, 1976, p. 15.
32. Arthur C. Clarke, *Profiles of the Future* (Bantam Books, 1967), p. 212.
33. *Scientific American,* Volume 213, Number 3, September, 1966, p. 246.
34. *IR,* May, 1976, p. 17; *SN,* Volume 109, February 28, 1976, p. 133.
35. *The National Enquirer,* June 1, 1976, p. 14.
36. Clarke, *Profiles of the Future,* p. 225.

Chapter 4

1. See reference 27 in chapter 2.
2. *CEN,* December 18, 1967, p. 21.
3. *SN,* Volume 102, July 22, 1972, p. 53.
4. The Oak Ridge, Tennessee, *Oak Ridger,* August 17, 1972, p. 1.
5. *SN,* Volume 103, June 30, 1973, p. 419.
6. *SN,* Volume 102, August 19, 1972, p. 125.
7. *SN,* Volume 111, January 1, 1977, p. 5.
8. *Newsweek,* February 3, 1975, p. 39.
9. *Argosy,* August, 1971, p. 46.
10. *Newsweek,* May 13, 1974, p. 122.
11. *Scientific American,* December 30, 1971, p. 30.
12. *SN,* Volume 106, December 7, 1974, p. 360.
13. *Newsweek,* March 11, 1974, p. 52.
14. *SN,* Volume 105, April 20, 1974, p. 184.
15. *SN,* Volume 108, July 19, 1975, p. 36.
16. *SN,* Volume 109, May 22, 1976, p. 314.
17. Clarke, *Profiles of the Future,* p. 235.
18. *CEN,* February 12, 1962, p. 138; February 19, 1962, p. 104; July 24, 1972, p. 13.
19. *True* magazine, June, 1971, p. 13.
20. *SN,* Volume 102, December 23, 1972, p. 413.
21. *Newsweek,* April 16, 1973, p. 58; *SN,* Volume 100, July 10, 1971, p. 27.
22. *SN,* Volume 108, October 11, 1975, p. 231.
23. *SN,* Volume 111, January 8, 1977, p. 27.
24. Hans Selye, *Calciphylaxis* (University of Chicago Press, 1962).
25. *Family Weekly* magazine, March 28, 1976, p. 19.
27. *Chemistry* magazine, Volume 37, Number 6, June, 1964, p. 7; *Finska Kemists Medd.,* 80, Number 2, 1971, p. 23.
28. *The National Tattler,* December 22, 1974, p. 8.

Chapter 5

1. *Newsweek,* July 15, 1975, p. 72; *SN,* Volume 108, November 8, 1975, p. 289; *Newsweek,* March 22, 1976, p. 59.
2. Louise B. Young, ed., *The Mystery of Matter* (Oxford Press, 1965), p. 321.
3. *SN,* Volume 101, June 3, 1972, p. 367.
4. *Technology Review,* May, 1971, p. 4; *Science Digest,* September, 1971, p. 24.
5. *CEN,* December 6, 1971, p. 24.
6. *CEN,* June 22, 1970, p. 80.
7. *Newsweek,* December 14, 1970, p. 118; *SN,* Volume 99, December 5, 1970, p. 429; *SN,* Volume 106, April 8, 1972, p. 231.
8. A. Lee McAlester, *The History of Life* (Prentice-Hall, Inc., 1968), p. 7; *SN,* Volume 105, January 26, 1974, p. 54.
9. *SN,* Volume 104, August 18–25, 1973, p. 102; Volume 108, July 19, 1975, p. 38.
10. *The Christian Science Monitor,* January 13, 1972, p. 7.
11. *Scientific Research* magazine, September 30, 1968, p. 27.

12. Gordon Rattray Taylor, *The Biological Time Bomb* (The World Publishing Company, 1968), p. 186.

Concluding Comments

The introductory quote by Arthur C. Clarke can be found in his book, *Profiles of the Future* (Bantam Books, 1967), pp. xiv and 1.

Part II

The introductory quote by Jonas Salk can be found in "The New Man—What Will He Be Like?," Part 4, Albert Rosenfield, *Look* magazine, October 1, 1965, p. 96.

Chapter 1

1. G. Eastman, "Scientism in Science Education," *The Science Teacher,* Volume 36, April, 1969, p. 19.
2. Quoted in *New York Times,* March 26, 1967; Also see "How Technology Will Shape the Future" by Emmanuel Mesthenes, *Science,* Volume 161, July 12, 1968, p. 135.
3. *Science Journal,* October, 1969, p. 39.
4. Edwin P. Booth, editor, *Religion Ponders Science* (Appleton-Century-Crofts, 1964), p. 144.
5. Rene Dubos, *Reason Awake* (Columbia University Press, 1970), p. 11.
6. See Eugene Wigner, *Physical Science and Human Values* (Princeton University Press, 1946).
7. See W. T. Stace, *Religion and the Modern Mind* (Lippincott, 1960).
8. *Science,* August 21, 1970, p. 721.
9. Kenneth E. Watt, *The Titanic Effect* (E. P. Dutton Co., 1974), p. 11.
10. Raymond Bauer, *Second Order Consequences: A Methodological Essay on the Impact of Technology* (MIT Press, 1969); quoted in *The Futurist,* December, 1971, p. 240.
11. *CEN,* January 10, 1972, p. 27.
12. *CEN,* September 27, 1976, p. 4; *SN,* Volume 110, September 25, 1976, p. 198.
13. James Reston, editorial in *New York Times,* May 29, 1964.
14. *The Futurist,* Volume 9, Number 5, October, 1975, p. 266.
15. *Midnight* magazine, October 14, 1974, p. 16.

Chapter 2

1. Ian G. Barbour, *Chemistry and the Scientist* (Association Press, 1960), p. 17.
2. Paul R. and Ann H. Ehrlich, *Population, Resources, Environment* (W. H. Freeman and Co., 1970).
3. *SN,* Volume 110, August 14, 1976, p. 104.
4. *SN,* Volume 106, July 6, 1974, p. 6.
5. *CEN,* May 13, 1974, p. 44; *SN*, Volume 110, October 9, 1976, p. 231.
6. *The National Enquirer,* May 17, 1977, p. 2.
7. *Bulletin of Atomic Scientists,* February, 1972, p. 11.
8. *Newsweek,* January 6, 1975, p. 36; *The Futurist,* October, 1976, p. 288.
9. *SN,* Volume 105, June 8, 1974, p. 368.
10. *SN,* Volume 108, October 4, 1975, p. 217.

11. *Science Digest,* August, 1971, p. 58.
12. *The Futurist,* Volume V, Number 5, October, 1971, p. 190.
13. *SN,* Volume 95, April 12, 1969, p. 356.
14. *CEN,* April 7, 1975, p. 24.
15. *National Geographic Magazine,* September, 1976, p. 397.
16. *The National Enquirer,* August 26, 1975, p. 24.
17. *CEN,* April 14, 1975, p. 6.
18. *CEN,* March 15, 1976, p. 7.
19. See reference 10 above.
20. *CEN,* August 23, 1976, p. 27.
21. *Family Weekly* magazine, August 26, 1975, p. 24.
22. *Newsweek,* June 3, 1974, p. 59; *SN,* Volume 105, June 29, 1974, p. 418.
23. *CEN,* August 11, 1975, p. 19.
24. *Bulletin of Atomic Scientists,* June, 1975, p. 23.
25. *IR,* October, 1975, p. 20.
26. *IR,* November, 1975, p. 51.
27. *IR,* December, 1974, p. 30.
28. According to Dr. Lorne D. Proctor of the Edsel B. Ford Institute for Medical Research, in March, 1966 newspaper accounts.
29. *CEN,* June 14, 1976, p. 21.
30. *SN,* Volume 105, March 9, 1974, p. 164.
31. *SN,* Volume 100, December 18, 1971, p. 407.
32. *SN,* Volume 105, June 1, 1974, p. 348.
33. *SN,* Volume 111, March 19, 1977, p. 190.
34. *Plain Truth* magazine, June, 1976, p. 31; *Bulletin of Atomic Scientists,* February, 1977, p. 17.
35. *Bulletin of Atomic Scientists,* December, 1972, p. 20.
36. *World Health* magazine, December, 1975, p. 4.
37. *Midnight* magazine, October 29, 1973, p. 8.
38. *SN,* Volume 95, January 11, 1969, p. 33.
39. *Psychology Today,* October, 1973, p. 59.
40. *Newsweek,* April 8, 1974, p. 52.
41. *SN,* Volume 103, February 3, 1973, p. 76.
42. *SN,* Volume 108, November 29, 1975, p. 347.
43. *CEN,* May 8, 1972, p. 4; *SN,* Volume 110, September 4, 1976, p. 153.

Chapter 3

1. *SN,* Volume 110, February 5, 1977, p. 89.
2. *Newsweek,* September 22, 1975, p. 87.
3. *SN,* Volume 101, May 6, 1972, p. 295; Volume 108, June 17, 1972, p. 392.
4. *The Futurist,* April, 1975, p. 75; *SN,* Volume 108, August 16, 1975, p. 106; *SN,* Volume 111, January 22, 1977, p. 59.
5. *Science,* Volume 174, November 19, 1971, p. 779.
6. *Newsweek,* September 22, 1975, p. 48.
7. *The Lutheran Standard,* February 3, 1976, p. 12.
8. *Perspectives in Biological Medicine,* Volume 14, 1970, p. 98.

9. *Science Digest,* August, 1971, p. 17.
10. *Midnight* magazine, September 16, 1974, p. 15.
11. *Plain Truth* magazine, March, 1977, p. 40; A letter to the editor of the *Bulletin of Atomic Scientists* by Douglas DeNike, October, 1976, p. 4.
12. Quoted in *Plain Truth,* March, 1977, p. 41.
13. *CEN*, March 24, 1977, p. 18.
14. *Plain Truth,* March, 1977, p. 40.
15. Gordon Rattray Taylor, *The Biological Time Bomb* (The World Publishing Co., 1968), p. 184.
16. *SN,* Volume 105, April 20, 1974, p. 256.
17. *Scientific American,* September, 1966, p. 322.
18. *SN,* Volume 109, February 28, 1976, p. 133; *TV Guide,* August 28, 1976, p. 22; *The National Enquirer,* June 1, 1976, p. 61.
19. *Motive* magazine, November, 1967, p. 37.
20. *Journal American Medical Association,* Volume 227, Number 9, March 4, 1974, p. 1027; *SN,* Volume 106, August 24 and 31, 1974, p. 119.
21. *The National Enquirer,* March 22, 1977, p. 22.
22. *SN,* Volume 101, May 27, 1972, p. 348; Volume 110, September 18, 1976, p. 183.
23. *Plain Truth,* June 21, 1975, p. 14.

Chapter 4

1. Leo J. and George Muir, Jr., *Muir's Thesaurus of Truths* (Deseret News Press, 1937), p. 391.
2. Floyd W. Matson, *The Idea of Man* (Delacorte Press, 1976), page xiii.
3. Edwin P. Booth, editor, *Religion Ponders Science* (Appleton-Century-Crofts, 1964), p. 25.
4. *Bulletin of Atomic Scientists,* Volume 28, Number 10, December, 1972, p. 20.
5. Teacher's Guide to "The Ascent of Man," by Jacob Bronowski, Mobil Oil Corporation, p. 25.
6. *Time* magazine, April 19, 1971, p. 47.
7. *Ibid.,* p. 33.
8. *Lay Sermons,* quoted in *Bartlett's Familiar Quotations* (Pocket Books, Inc., 1965), p. 187.
9. Dennis and Donnella Meadows, editors, *Toward Global Equilibrium: Collected Papers* (Wright-Allen Press, Inc., 1973), p. 346.
10. *Ibid.,* p. 345.
11. Anthony Standen, *Science Is a Sacred Cow* (E. P. Dutton Co., 1950), p. 13.
12. *Ibid.*, p. 13.
13. See *Mysteries of Science* by John Rowland (Werner-Laurie Ltd., 1955).
14. *CEN,* August 21, 1967, p. 54.
15. *Science Is a Sacred Cow,* p. 52.
16. *Scientific American,* January, 1955, p. 437.
17. *Science,* Volume 185, July 26, 1974, p. 303.
18. *SN,* Volume 107, March 8, 1975, p. 148; Volume 107, March 22, 1975, pp. 187, 194; Volume 107, June 7, 1975, p. 366; Volume 110, July 3, 1976, p. 3.
19. *SN,* Volume 111, January 29, 1977, p. 69.
20. *The National Enquirer,* February 23, 1977, p. 6.

21. Quoted in *Plain Truth,* March, 1977, p. 40.
22. *Bulletin of Atomic Scientists,* May, 1977, p. 10.
23. *CEN,* February 17, 1977, p. 6.
24. Ian G. Barbour, *Christianity and the Scientist* (Association Press, 1960), p. 55.
25. *Religion Ponders Science,* p. 28.
26. *Science,* Volume 174, November 19, 1971, p. 787.
27. From an article by Hans Jonas, which appeared in a Courses by Newspaper Series titled "Moral Choices in Contemporary Society," in April of 1977.
28. Quoted in *Plain Truth,* March, 1977, p. 39.
29. Quoted in *Plain Truth,* March, 1977, p. 39.
30. Quoted in *Wednesday Night at the Lab,* edited by Kenneth L. Rinehart, Jr., William O. McClure, and Theodore L. Brown (Harper and Row, 1973), p. 134.
31. Ibid., p. 134.
32. From *The New Prometheans* by John S. Lambert (Harper and Row, 1973); quoted in "Prometheans and Epimetheans" by Frank Snowden Hopkins, *The Futurist,* June, 1974, p. 134.

Chapter 5

1. *The National Enquirer,* Volume 46, Number 22, January 30, 1972, p. 2.
2. *Science,* Volume 174, November 19, 1971, p. 786.
3. *Bulletin of Atomic Scientists,* Volume 32, Number 10, December, 1976, p. 9.
4. Choruses from "The Rock," in *Collected Poems, 1909–1935,* (Harcourt, Brace, and Co., 1936), p. 179.
5. *Plain Truth* magazine, October 18, 1975, p. 14.
6. Ibid., p. 14.
7. Robert L. Wolke, editor, *Impact: Science on Society* (W. B. Saunders Co., 1975), p. 184; reprinted from *Center* magazine.
8. "Growth and Education: A Strategic Report to the Rockefeller Brothers Fund On the Implications of Growth Policy for Postsecondary Education," Western Interstate Commission for Higher Education, Post Office Drawer "P," Boulder, Colorado, 80302, December, 1974.
9. John Clover Monsma, editor, *Science and Religion* (G. P. Putnam's Sons, 1962), p. 22.
10. Loren Eiseley, *The Immense Journey* (Random House, 1957), p. 210.

Concluding Comments

1. John Clover Monsma, editor, *Science and Religion* (G. P. Putnam's Sons, 1962), p. 17.
2. Leroy Augenstein, *Come Let Us Play God* (Harper and Row, 1969), p. 146.
3. *J. Cent. Cong. Amer. Rabbis,* January, 1968, p. 27.
4. *Science,* Volume 174, November 19, 1971, p. 785.
5. Quoted in *CEN,* August 23, 1976, p. 41.
6. *Bartlett's Familiar Quotations* (Pocket Books, Inc., 1965), p. 146.